THE
HAMLYN
DICTIONARY
—of—
GEOGRAPHY

THE
HAMLYN
DICTIONARY
of
GEOGRAPHY

Roy Woodcock

HAMLYN

First Published 1989 by
The Hamlyn Publishing Group Limited
Michelin House, 81 Fulham Road
London SW3 6RB

Copyright © The Hamlyn Publishing Group Limited 1989

ISBN 0 600 56519 X

Typeset by J&L Composition Ltd, Filey, North Yorkshire

Printed in Great Britain by
William Collins and Sons

INTRODUCTION

Geography as a subject has undergone profound changes over the past 20 years and now embraces the whole environment, both natural and man-made. THE HAMLYN DICTIONARY OF GEOGRAPHY takes due account of these changes and offers more than just a practical guide to the science of the earth's surface: it presents an extensive and up-to-date survey covering both physical and human geography. It has more than 1000 entries, ranging in scope from *ablation* and *age sex pyramid* to *urban morphology* and *zero population growth*. Clear definitions are accompanied by a careful system of cross-referencing which guides the reader from one related entry to another. The text is enhanced by a useful selection of line illustrations. The DICTIONARY will be of great value to the student of geography (at both G.C.S.E. and A-level) and to anyone who is interested in this fascinating subject.

NOTE: An asterisk (*) following the entry headword indicates that the entry is accompanied by an illustration.

A

aa The Hawaiian name for blocky lava, a type of lava flow which has solidified into a blocky mass, rather like cinders in appearance. Gases would have escaped from the molten lava in a fairly violent fashion, making the rock jagged and angular. Examples are found on Mauna Loa in the Hawaiian Islands, Mt Etna in Sicily and Mt Hekla in Iceland. See also **pahoehoe**.

ablation 1 Loss of ice in a glacier due to evaporation and melting. The process includes melting at the top of the ice because of sunshine, and around the edges of the ice because of rain water or melt water. 2 The loss of ice at the end of a glacier where an iceberg may break off.

aborigines The native inhabitants of any country, now normally used to refer to the original inhabitants of Australia only. Some Australian aborigines still live in very isolated parts of Northern Australia, but many now work on farms or live in the towns. They formerly lived by hunting and gathering, using boomerangs to kill birds and animals, and were skilful at finding food in a very difficult environment.

abrasion The wearing away of rocks by the action of wind, water, or ice carrying particles of dust and sand; a mechanical progress, which acts rather like sandpapering. Abrasion in rivers will erode the banks and may form potholes on the bed. Abrasion is most active where there is no vegetation to protect the rocks and soil, especially in deserts. Mushroom-shaped rocks have been formed by abrasion, and telegraph poles have been worn through by wind-blown sand. Abrasion is effective along coastlines, too, where it will wear away at the cliffs as well as the shore. See also **wave-cut platform**.

abrasion platform see **wave-cut platform**

absentee landowner A farm owner who lives in a town and not on his farm, leaving a manager to run his estate though the owner will collect all the profits.

absolute humidity The amount of water vapour present in the air at any given time. It is measured in grams per cubic metre. The amount of water vapour which the air can hold will depend on its pressure and temperature. Warm air can hold much more water vapour than cold air. For example, air at 10°C can hold 9.4 grams per cubic metre, whereas at 20°C it can hold 17.1 grams. When the air contains the maximum amount it can hold, it is said to be saturated and to have reached **dewpoint**, that is, the temperature at which dew begins to be deposited. There is no specific temperature for dewpoint, but the relative humidity will be 100%.

accessibility The term used in transport studies, to indicate the ease or difficulty in getting from one place to another. Accessibility may refer to cars and roads, or to public transport on roads or railways. Central locations in towns are often very accessible because they have many routes. Peripheral locations may have fewer routes but may suffer less from traffic congestion. Some towns have great accessibility because of high-speed trains, and for longer distances air travel will be important.

acid lava Lava which is viscous and slow-flowing, rather like thick porridge. It contains a high proportion of silica and usually solidifies quite quickly. Acidic volcanoes have much steeper cross sections than basic volcanoes. One of the most famous of the acid lava volcanoes is Mt Pelée in Martinique, which had a disastrous eruption in 1902. See also **basic lava**.

acid rain Rain which contains pollutants, especially sulphur and nitrogen oxides. The pollutants result from the burning of fossil fuels, especially in thermal power stations, but they can also be caused by factories and car exhausts. Once they are in the atmosphere, pollutants can be washed down to earth by rainfall as dilute forms of sulphuric and nitric acid. Although dilute, they are powerful enough to poison lakes and rivers, kill trees and speed up the erosion of buildings. Most industrial countries now have strict regulations to control the amount of pollution which is sent into the atmosphere. Winds often blow the pollution away from the countries which caused it; thus Norway and Sweden frequently suffer from Russian, German and British pollution.

acid rock Any of the igneous rocks which contain a high proportion of silica (more than 66%). The main silica mineral is quartz, but some felspars are also quite rich in silica. Acid igneous rocks are usually light in colour. The most common is granite. When they are weathered to form soil, they are generally infertile.

acid soil Soil which is short of bases and has pH of 6 or less. In wet and cool areas, water leaches out all the soluble bases, especially calcium, to form acid soil. See also **podsol**.

acre An imperial unit of area covering 4840 sq yd. It was based on a furlong (220 yd) multiplied by a chain (22 yd): 640 ac equal 1 sq mi, and 1 ac equals 0.4047 hectares. One acre is about half the size of a football pitch.

acre foot (in irrigation) The amount of water required to cover an acre to a depth of one foot, 43,560 cu ft.

adiabatic Denoting changes in the pressure and temperature of a parcel of air, when heat from outside is neither added nor removed. The air will be either expanding or contracting, thus producing changes of pressure and temperature. See also **dry adiabatic lapse rate, saturated adiabatic lapse rate**.

adit A type of mining in which the miners dig in horizontally from the side of a valley. It was used extensively in the coal mines of South Wales. It is cheaper than having to sink a deep shaft, but can be used only when erosion has exposed the coal seam.

adobe Mud which has been dried in the sun and used as bricks. Adobe bricks are used in the southern United States and Mexico, and also in Argentina and other parts of South America. Similar mud bricks, often containing straw to make them stronger, are used in many parts of Africa. Heavy rain will gradually soften them, and so adobe houses tend not to last for a long time.

adret The south-facing slope, or sunny side, of a valley, especially in the French Alps. It is the side of the valley on which most houses will be built, and also where the best farming can take place. See also **ubac**.

aerobic A term denoting organisms which live in a location with free oxygen. See also **anaerobic**.

afforestation The planting of trees, generally in an area which has not had trees in the past. For example, the Forestry Commission has planted trees on moorland areas of Wales and Scotland, and the Soviet government has planted millions of trees on the dry steppes of Kazakhstan. Trees may be planted in order to produce wood or pulp some time in the future, or as windbreaks, or to hold the soil together and reduce soil erosion. In some Third World countries they are now being planted to provide firewood.

age sex pyramid* A frequency distribution histogram which shows the age and sex composition of a population. The horizontal bars are usually indicated as percentages of the total population, but can be drawn as actual numbers, or as a percentage of just the male or just the female populations. Each horizontal bar can represent any number of years, but it is common to draw five-year groups (called **cohorts**). In this case the lowest bar would represent all children aged 0–4, and the next bar would show the 5–9 cohort, and the next would be 10–14 and so on. It is normal to put the males on the left of the central axis, and the females on the right. The shape of the pyramid will vary according to the population composition of the town or country which is the subject of the histogram. Also called **population pyramid**.

agglomerate A type of rock which has been formed by volcanic activity. It consists of coarse blocks of rock or lava which have been thrown out by a volcanic eruption and then cemented together in volcanic ash.

agglomeration 1 A group of houses or settlements which have spread and grown together to merge with several neighbouring settlements. 2 (In industrial geography) a group of industries which are close to one another and possibly related in what they produce or in the materials they use. An industrial agglomeration often occurs in coalmining areas, when steel works, then engineering works, and then chemicals all find that the best location for them is on the coalfield, as for example, in south Wales, northeast England, Nord in France and the Sambre-Meuse valley in Belgium. There are no industrial agglomerations in countries such as Sweden or Switzerland where hydroelectricity rather than coal was the main source of energy when industries were being developed.

agricultural revolution The period during the 19th century when new machines and farming methods enabled extra food to be produced and

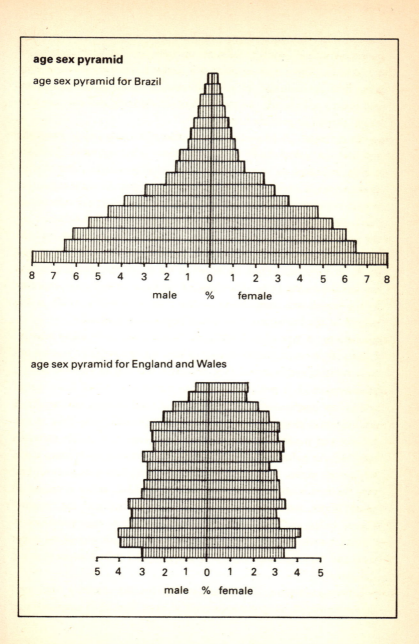

age sex pyramid

age sex pyramid for Brazil

8 7 6 5 4 3 2 1 0 1 2 3 4 5 6 7 8

male   %   female

age sex pyramid for England and Wales

5 4 3 2 1 0 1 2 3 4 5

male   %   female

new areas to be turned into productive land. It was not really a revolution, but merely part of the continuing change which had been affecting farming for centuries. More dramatic changes have probably occurred during the last 30 or 40 years, with many new fertilizers and pesticides, as well as new machines such as combine harvesters. See also **green revolution**.

agriculture The cultivation of the soil and the management of the natural landscape in order to grow crops, including both food and industrial crops such as cotton or rubber. There are many different types of agriculture, and the term is frequently used to include pastoral farming. See also **commercial agriculture, extensive farming, intensive farming, subsistence farming**.

A horizon The top layer of soil which contains fine particles of soil together with the humus. There is much mineral and organic material, and most of the plant roots are found in this layer. It is the first part of any soil to be washed or blown away if erosion is taking place. Resting on top of the A horizon there may be an **O layer**, which is where organic material accumulates. The O layer may consist of L, which is the litter layer; F, which is the partly decomposed or fermented layer; and H, which is the well-decomposed or humified layer. See also **B horizon, C horizon**.

air frost see frost

air mass A large body of air covering hundreds of square kilometres. When a northerly air mass travels over Britain it will have come from the Arctic and will bring cold weather. Air masses are named after the region they have come from. A polar air mass will have travelled from north of the Arctic Circle, and a tropical continental air mass will have come from the Sahara, or parts of the Arabian peninsula, or Australia. All air masses gradually change as they travel, becoming wetter if they move over sea, or drier if they travel over land. If they are moving towards the equator they will gradually get warmer, but if they are travelling towards the Arctic (or Antarctic) they will become colder.

air stream An air movement which is smaller than an air mass, and is simply a moving current or channel of air; a large wind.

alfalfa A leguminous plant, *Medicago sativa*, similar to clover. It has long roots and can survive spells of dry weather. It is a nitrogenous plant, which means it takes nitrogen from the air and stores it in its roots; this is then released into the soil to restore fertility. Alfalfa is a good fodder crop for cattle and is widely grown in the United States, Argentina, and many other countries.

alluvium Fine mud or silt deposited by rivers. It accumulates on river beds but can be spread over large areas of a river valley at times of flood. All rivers have some alluvium deposits, but large rivers such as the Mississippi or Amazon have alluvial plains which are several kilometres wide. Alluvial deposits at the foot of steep mountains often accumulate in a triangular or cone shape, rather like an inland delta. Such deposits are called **alluvial fans** or **alluvial cones**. Examples of these can be seen in many valleys in the Alps but smaller ones also occur in the Pennines and the Lake District.

alp 1 An area of grassland vegetation high up on the side of any mountain range. The alpine pasture is used by grazing animals in the summer but is snow-covered in winter. On the alp there will probably be huts or chalets, which are used by local people in the summer while looking after the cattle. In Switzerland and other countries some of the alps have become important locations for skiing. **2 The Alps** A mountain range which extends from France through Switzerland, Italy and Germany to Austria and Yugoslavia.

alpine folding The most recent large-scale example of folding, which occurred when the Alps, Himalayas, Rockies, Andes, and other large mountain ranges were formed. See also **fold mountains**.

anabatic see **katabatic**

anaerobic A term denoting organisms living in a location without oxygen. This would be in a very wet place, such as a peat bog. In anaerobic conditions dead vegetation decays very slowly and may possibly form coal or peat. See also **aerobic**.

anemometer An instrument for measuring wind speed. It consists of three cups which are blown round by the wind, and the speed of the wind can be read off the dial. More sophisticated instruments keep a written record of wind speed and can record wind direction as well. An anemometer should be located in an open space, well away from buildings and trees, and about 10 m above ground level if possible.

aneroid barometer An instrument for measuring atmospheric pressure. Less accurate than a mercury barometer, it consists of a box with most of the air removed, the sides of which move in and out with changes of pressure. The needle recording the pressure changes can be attached to a pen which writes the pressure readings on to a piece of paper attached to a revolving cylinder. In this way a written record of a week's pressure can be obtained. See also **barograph**.

anthracite A hard type of coal which burns with little visible flame and leaves little ash. It consists of more than 90% carbon and can be used as smokeless fuel. Large deposits have been found in south Wales and in Pennsylvania in the United States. See also **bituminous coal, brown coal**.

anticline The arch of upfold in layers of rock strata. It is the result of compression, which caused the rock strata to bend. See also **syncline**.

anticyclone An area of high pressure where air is subsiding, causing dry weather. Winds are normally light and the weather is settled. Major high pressure regions are found in the horse latitudes, about 20° to 30° north and south of the equator. There are also large anticyclones in the Arctic and Antarctic. High pressure over Britain in summer will bring hot and sunny weather because the air will have come from the Sahara and the Mediterranean regions, but in winter the high pressure may have come from the Arctic and thus result in cold weather. Also called **high**.

appropriate technology A term used to denote small-scale or simple tools or machines which are more useful to people in poor areas than the kind of high technology used in the developed world. For example, small hand-held ploughs or diggers are likely to be more appropriate for many farmers than a large tractor or combine harvester; a small loom may be more use in an African village than a large electric-powered machine. Such machinery is easily repaired and should not be dependent on foreign experts or expensive parts. Also called **intermediate technology**.

aquifer A layer of rock which can hold large quantities of water in the pore spaces. Water may percolate along an aquifer, following the gradient of the stratum. An aquifer is generally located between two impervious layers.

arable A term denoting the type of farming in which crops are grown. Arable farmland is generally suitable for ploughing.

Archimedes' screw A device for raising water consisting of a screw inside a tube; as the screw is turned water passes up the thread. The lower end of the tube is placed in a river or canal, and water flows out at the upper end. It is used for irrigation in the Nile valley and elsewhere in Africa.

archipelago A group of islands. It may be a broken chain of islands where a range of mountains has been submerged by the sea, a collection of coral islands, or a group of volcanic islands, where submarine eruptions have built up islands from the ocean floor. Examples include the Aleutian Islands and the Bahamas.

Arctic Circle The line of latitude at 66° 30'N. All points along this latitude have a twenty-four hour period of continuous daylight in summer and a twenty-four hour period of continuous darkness in winter. Moving in a northerly direction from the Arctic Circle, the periods of continuous daylight or darkness become progressively longer.

area of outstanding natural beauty (AONB) An attractive scenic area in England or Wales which has been designated as such. Used as leisure and tourist locations, they are controlled by local authorities. There are over thirty AONBs, including parts of the South Downs and the Cotswolds.

arenaceous see **rudaceous rock**

arête A narrow knife-edged ridge which separates two corries, or a corrie and a U-shaped valley, formed by glacial erosion and freeze-thaw activity. In the Alps and Himalayas some arêtes are only a few centimetres wide and may have steep slopes which drop more than 300 m. In Britain arêtes are often wider and not quite as high; Striding Edge on Helvellyn is one of the steepest in England, but there are larger examples in the Cuillin Hills of Skye.

argillaceous see **rudaceous rock**

aridity A term denoting dryness because of lack of rainfall. An arid area receives less than 250 mm of rainfall per annum, and only specialized drought-resistant plants can survive. There is much bare rock in arid regions, with cactus, spiny bushes, or tussocks of grass forming a sparse cover of vegetation.

arithmetic growth A growth rate with a regular progression; for example, 2, 4, 6, 8, 10. Thomas Malthus thought that food production would increase at an arithmetic rate, while population would grow at a geometric rate.

Armorican see **Hercynian**

artesian A term denoting water which has moved from its place of origin by travelling underground, usually by means of percolating along an aquifer.

artesian basin A synclinal structure with a permeable stratum between two impermeable strata. The largest artesian basin in the world is in Australia, and many farmers in southwest Queensland depend on artesian water for their cattle and their sheep. The water which comes up their boreholes probably fell as rain on the Great Dividing Range several hundred kilometres to the east and northeast. In some cases, where the water-table of the nearby hills is higher than the top of the well which is tapping the artesian water, the water may squirt out under its own pressure, possibly as a fountain.

arroyo see **wadi**

artificial fibre see **synthetic fibre**

ash The fine particles of rock thrown out by a volcanic eruption. If the ash is not blown very high, it will fall quickly and form a depositional area downwind of the volcano. If it is blown thousands of feet into the atmosphere, as at Krakatoa in 1883, some of the dust may circle the earth for several months. After Mt St Helena erupted in 1980, a cloud of ash drifted across the United States and the Atlantic Ocean.

aspect The direction a location is facing. The aspect is sometimes important for crop growing, as many plants need to face south in the northern hemisphere in order to get enough sunshine to ripen; for example, grapes. It may also be important for people, for example, in order to avoid cold northerly winds.

assembly industry An industry engaged in gathering components and putting them together to make a finished product. The components may be made by several different firms. The motor car industry assembles the body, seats, windscreen, engine, etc. as the car slowly moves along a conveyor belt.

asymmetric fold A lop-sided fold, caused by greater pressure from one side than the other. Most folds are asymmetric. A good example are the hills which run from west to east across the Isle of Wight.

Atlantic coast see **discordant coastline**

atmosphere The layer of air which surrounds the earth. It consists of the permanent gases, including oxygen (21%), nitrogen (78%), argon, and helium, and the variable gases, such as ozone, which occurs mostly over 30 to 60 km above the earth's surface, water vapour, which is mostly below 15 km, sulphur dioxide and carbon dioxide. The atmosphere is about 15 km thick at the equator, but becomes thinner towards the poles. As the height above the earth's surface increases the density of the atmosphere becomes thinner, or more rare. However, the relative proportions of the gases remain fairly constant, although the amount of water vapour does decrease with height. Most of the weather phenomena occur in the lowest parts of the atmosphere. See also **stratosphere, tropopause, troposphere**.

attrition A process of erosion in which the particles or tools of erosion erode one another. For example, the sand particles in a river's load may gradually erode one another; small particles carried by the wind will also suffer from attrition, as will the pebbles on a beach. Attrition gradually reduces the size of the particles, making them smooth and rounded.

automation The use of machines to replace manpower. Automation is found in many manufacturing industries, such as motor cars or textiles, but only in wealthy countries, since it is very capital intensive.

autumnal equinox see **equinox**

avalanche A large amount of snow and ice or rock falling suddenly down a mountainside. Avalanches occur when the temperature rises quickly causing the snow to melt; they can also be started by a sudden noise.

azonal soil A soil associated with a particular environment, such as a scree slope, sand-dune, or glacial moraine, on which little real soil development has taken place and there are no true soil horizons.

Azores high A region of high pressure which occasionally affects the British Isles, bringing hot, sunny and settled weather. In some years there may be Azores high pressure over Britain for several weeks, but in other years there may be none at all. The Azores high is part of the tropical high pressure which causes the Sahara desert and the hot dry summers over the Mediterranean.

B

backing A meteorological term denoting the anti-clockwise change of direction of a wind, such as south-westerly to southerly to south-easterly. See also **veering**.

backwash The downshore movement of water on a beach. When a wave breaks, the **swash** takes the water up the beach; the backwash brings it down again. The backwash may be on the surface or in the sand or shingle on the beach. It normally runs down the steepest slope, which is usually at right angles to the sea.

badlands An eroded landscape in an arid or semi-arid area. Because of the lack of vegetation, the rain water runs off very quickly and erodes any soft or exposed rocks, forming a steep and furrowed landscape which is impressive but quite useless for agriculture. The most noted badlands are in South Dakota and Nebraska in the United States.

barchan Also **barkhan** 1 A crescent-shaped sand-dune with the horns pointing downwind, found in most of the large sand deserts of the world, such as southern California, parts of Libya, and Turkestan. Barchans form in areas where the wind direction is fairly constant. Wind blows over the crest and round the edges of the horns, so that the entire dune gradually moves downwind. 2 A similar feature of dry powdery snow.

bar graph* A graph in which columns are used to represent the values of the information recorded. Several bars can be drawn close together to enable comparisons to be made. A simple bar shows a total value, while a compound bar can be sub-divided to show constituents as well as total values. The divided bar graph can be used to show the kind of information which is often illustrated by pie charts, and the divided bars are easier to draw than circles. The bars may be drawn horizontally or vertically. Horizontal bars can be placed one on top of the other to form a pyramid, as in a population pyramid; alternatively, they can be

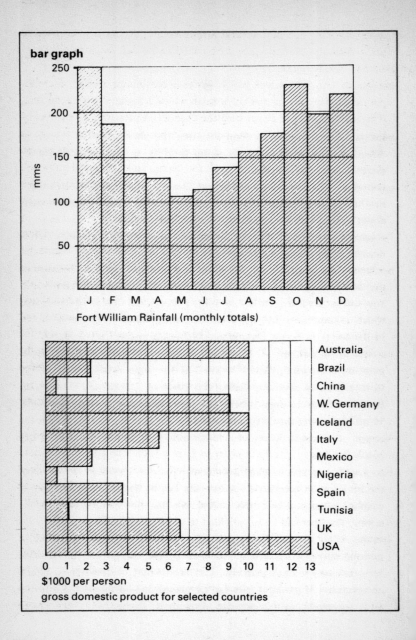

bar graph

Fort William Rainfall (monthly totals)

gross domestic product for selected countries
$1000 per person

Australia
Brazil
China
W. Germany
Iceland
Italy
Mexico
Nigeria
Spain
Tunisia
UK
USA

drawn side by side in a vertical arrangement, such as is often used to show monthly rainfall totals. Also called **battleship diagram**.

barkhan see **barchan**

barograph An instrument which writes a continuous record of atmospheric pressure. It consists of a small aneroid barometer with an arm, which moves up and down with the changes in pressure. At the end of the arm is a nib, which rests on a clockwork drum bearing a piece of special recording paper; as the drum revolves, a thin line is drawn on the paper. At the end of each week the paper can be removed and a full record of the pressure changes can be seen. The barograph is more useful for recording the changes in pressure than for giving an accurate reading of atmospheric pressure at any particular time.

barometer An instrument for measuring atmospheric pressure. There are two main types: the mercury barometer and the aneroid barometer which is the less accurate, but is the type normally used in barographs and found in the home. The mercury barometer consists of a glass tube containing mercury, which is inverted over a small reservoir of mercury; as the pressure rises, the weight of the atmosphere forces mercury out of the reservoir and up the tube. The tube has a scale on it, so that the pressure can be read off at the top of the column of mercury. As the pressure falls, the column of mercury gradually moves down. A column of mercury 760 mm in length represents the average pressure at sea level, and so the mercury barometer has to be quite a long instrument in order to contain such a long tube.

barrage A large structure of concrete, or earth and rocks, designed to block the flow of a river in order to form a lake. The lake may be used as a source of water for irrigation, or may be designed to reduce flood danger. There is a barrage across the Rance estuary near St Malo in France, where a tidal power station has been erected. See also **dam**.

barrow see **tumulus**

basalt A black or dark coloured igneous rock which consists of fine-grained minerals because it was formed near the earth's surface from lava that cooled fairly quickly. The dark colour of basalt is caused by the presence of minerals which are basic rather than acidic in character, such as augite, hornblende, and mica. Basalt sometimes forms hexagonal columns when it cools from the liquid lava, as in Giant's Causeway in Northern Ireland and at Fingal's Cave on the island of Staffa. Basalt is formed by extrusive volcanic activity from lava that is fairly free-flowing and liquid. Flows of basalt sometimes spread over large areas of countryside; for example, in Iceland, the Deccan plateau in India and the Snake-Columbia plateau in the northwest United States.

basic lava Free-flowing lava, which spreads quickly across the country-side after an eruption. It is usually dark in colour because it contains basic ferro-magnesian minerals and is less than 50% silica. It may erupt through craters or from fissures, and forms basalt when it solidifies; fissure eruptions are likely to form plateaus. If a volcanic mountain forms it is likely to have gently sloping sides which spread over a very large area; for example, Mauna Loa in Hawaii. Basic lavas are common in Iceland on the mid-Atlantic Ridge; old basic lavas are found in the Antrim Mountains of Northern Ireland and on the island of Mull in western Scotland. Weathering gradually breaks up the lava to form soil which will be rich and productive, as in Java or on the slopes of Mt Etna. See also **acid lava**.

basin 1 An area in which the strata dip from the outside edges towards a central point, such as the Donets and Kuznetsk basins in the Soviet Union. 2 (a) The **catchment area** of a river. (b) An area of inland drainage in desert regions, where rivers which are unable to reach the sea either flow into lakes or evaporate in salt flats, such as Lake Eyre in Australia and the Great Salt Lake in the United States. 3 A large hollow or depression in the earth's surface, partly or wholly surrounded by higher ground. 4 A type of irrigation, found in India, Egypt, and elsewhere, in which embankments are constructed to prevent the escape of flood or irrigation water from a field. The water gradually sinks into the soil, thus enabling crops to be grown.

batholith A large area of igneous rock which cooled and solidified below the surface. Because of this it contains large crystals and will probably be granite. Batholiths can be seen at the surface when the overlying rocks have been worn away by erosion. As they are quite hard rocks they generally form high ground; for example, Dartmoor, Bodmin Moor, and sometimes they can form large mountain ranges, such as the Sierra Nevada in California.

battery farming see **factory farming**

battleship diagram see **bar graph**

bay An indentation or inlet in the shore or a sea or lake. It is generally formed as a result of varying rates of erosion, the softer rocks having worn away faster than neighbouring harder rocks. Bays may provide useful anchorage and are often the site of settlements.

bay bar A ridge of sediment deposited across the mouth of a bay and attached to the land at both ends. It consists of material which was deposited by a river or carried along the coast by longshore drift. If it completely blocks the bay from the sea, deposits will pile up on the landward side of the bar. This will gradually form a delta of reclaimed land. A good example of a bay bar can be seen at Loe in Cornwall. See also **spit**.

bauxite The ore from which aluminium is obtained. There are rich deposits of bauxite in Guinea, Jamaica, Brazil and Surinam.

beach The land on a shore between high-water mark and low-water mark. Some beaches consist of smoothed rocks which have been eroded by the sea; others may contain large and rugged rocks, also resulting from erosion. Many beaches contain material of deposition, such as shingle, sand, or mud. The constituents of a beach will depend on the coastal rock types, the amount of erosion, neighbouring rock types along the coast, and the amount of longshore drift; also, the presence of groynes may have a significant influence on the nature of a beach. Constructive action along a beach may be caused by the swash; destructive action is associated with the backwash.

bearing A compass point measured in degrees from 0 to 360. To work out a bearing start from north, which is 0° and 360°, and progress in a clockwise direction; east is 90°; south is 180°; and southwest is 225°.

Beaufort scale A scale of wind speed devised by Admiral Beaufort at the beginning of the 19th century. The Beaufort scale is used internationally.

Number	Wind	Average wind speed (km per hour)	Visible effects
0	calm	1	Smoke rises vertically.
1	light air	3	Wind direction shown by smoke.
2	light breeze	9	Wind felt on face; leaves rustle.
3	gentle breeze	16	Leaves and twigs in constant motion.
4	moderate breeze	24	Raises dust and loose paper.
5	fresh breeze	34	Small trees in leaf sway.
6	strong breeze	44	Large branches in motion.
7	moderate gale	56	Whole trees in motion.
8	fresh gale	68	Twigs break off trees.
9	strong gale	81	Slight structural damage.
10	full gale	95	Trees uprooted.
11	storm	110	Widespread damage.
12	hurricane	above 121	Devastation.

bedding plane The dividing line between two different strata of sedimentary rock. A bedding plane generally indicates a slight change of rock type; each stratum would have been deposited over a period of years, and then consolidated to form a layer of rock.

bed rock The solid rock which lies below the soil and the regolith layers. In soil profiles it will be the C or D horizon. In some places, such as the North American prairies or East Anglia, the bed rock is several metres below the surface, but in other places, especially on hillsides, it may be within a few centimetres of the surface.

bergschrund A large crevasse which forms in the ice at the upper end of a corrie, near the back wall. It is caused by the downhill movement of the ice, which gradually pulls away from the back wall, opening up a crack or crevasse.

beta index A quantitative method of analyzing a network of communications. It involves counting up the number of edges or arcs and nodes or vertices in a network of roads or railways. Dividing the number of nodes into the number of edges will give the beta index. The figure for one network can then be compared with the figure for another network in a precise and quantitative way. If there is one circuit in the network, the index will be one; two circuits will give an index of two, and so on. The greater the index, the better the connectivity of the network.

B horizon The layer beneath the A horizon in a fully formed soil profile. It has less weathered material and less humus than the A horizon, but often contains chemicals which have been washed down from above. See also **A horizon, C horizon**.

bid rent theory* The theory that rent or land values decrease with increasing distance from a central location. The centre may be a small village or the centre of a large town. Shop and office owners are prepared to pay higher rents for central and accessible locations, and therefore they tend to occupy the most expensive central areas. The ideas of bid rent occur in the von Thunen land-use model, and also in land-use models of urban areas. It is a sound general idea, but there are often exceptions caused by relief features, communications, or the idiosyncracies of human beings.

bilharziasia A debilitating disease of the blood in man and other animals. It is caused by the bilharzia worm, which develops in snails living in water, such as irrigation ditches, and enters the bloodstream of humans standing in the water through their feet; subsequently eggs are passed back into the water via faeces and urine. It is very common in the Nile valley and has become more widespread recently because of the increasing number of irrigation ditches.

biogas The gas produced by rotting plant or animal material, such as manure; methane. It is a renewable source of energy and has already been used on a small scale to drive motor engines in Britain and as a substitute for firewood in parts of India.

biogeography The study of the geographical distribution of plants and animals. Biogeography includes the study of ecology and ecosystems, soil studies and many other topics.

biological weathering see **weathering**

biomass The total organic matter of the plants and animals in a given

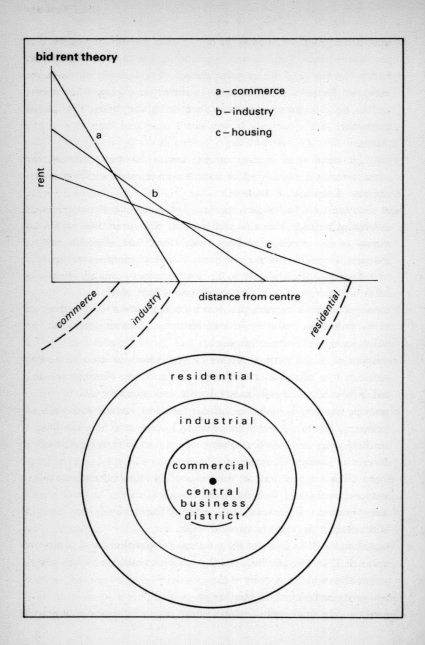

bid rent theory

a – commerce
b – industry
c – housing

rent

a

b

c

distance from centre

commerce

industry

residential

residential

industrial

commercial

central
business
district

area. It is usually expressed as the dried weight per unit of area, such as per square metre. Figures range from 45 kg per sq m in tropical rain forests, to only 0.02 kg in polar regions. The biomass of temperate deciduous forests is 30 kg per sq m; savanna areas average 6 kg; deserts are 0.7 kg and cultivated land is about 1 kg per sq m. The animal proportion of the biomass is very small compared with that of the plants.

biome A major type of environment, such as savanna, tropical rain forest, or tundra, considered as a whole with its plant and animal life.

birth rate The number of live births per year for every thousand people of the population under consideration. A birth rate of 40 per thousand means that for every thousand people in the population there will be an average of 40 live births each year. Some Third World countries such as Kenya still have high birth rates, but in European countries the figure is now much smaller, about 12 to 15. Two hundred years ago the figure for Britain was about 40. Also called **crude birth rate**.

bituminous coal A type of coal which contains about 80% carbon. It burns quite well, but gives off some smoke, and leaves some ash. The major users are domestic households and thermal power stations; for example, the Trent valley in northern England has several power stations which rely on bituminous coal from the Yorkshire, Nottinghamshire, and Derbyshire coalfields. See also **anthracite, brown coal**.

blackband iron ore A thin dark-coloured seam of iron ore found in coal measures. The blackband deposits found in such areas as south Wales, central Scotland and the Ruhr valley helped iron and steel industries to develop in those areas, although the deposits did not last very long. Once the blackband ore had been exhausted imported ore was used, and new steel works were sited near the ports.

Black Country The area to the west of Birmingham, but east of Wolverhampton, formerly associated with dirty, heavy industries. The local coal, iron and limestone gave rise to numerous iron and steel works near Dudley, Walsall, Halesowen, Smethwick, and other towns. Most of the mines and many of the industries have now closed, and the area is certainly much cleaner, although there are a number of small industries producing nuts, bolts, chains, and other small metal goods.

Because many of the raw materials have to be brought in, only small products are manufactured in the area, but the industries survive because of tradition and the skills of the available work force. This is an example of geographical inertia.

black earth see **chernozem**

block mountain A mountan or a plateau uplifted between two or more parallel faults. Good examples can be found in the Meseta in Spain, the Black Forest in Germany, the Vosges in France and the Grampians in Scotland. They are all steep sided because of the effects of the faulting. Also called **horst**.

blowhole* A hole in a cave or cliff through which water is forced, especially at high tide. Blowholes are formed by erosion, probably along a joint or line of weakness in the rocks, and commonly occur in cave roofs. Examples can be seen on many coastlines, such as Pembroke in south Wales or Caithness in Scotland.

blue-collar worker A manual worker; an employee who works in a potentially dirty environment and would therefore need to wear overalls or other protective clothing. See also **white-collar worker**.

bluff A steep slope at the side of a river formed by erosion on the outside of a meander. Bluffs may be quite small, but some are 100 m or more in height. Since they are steep, they are often covered with trees, and are not used for farming. Also called **river cliff**.

bora A cold wind which blows from the north or northeast into the northern part of the Adriatic. It is associated with the movement of a low pressure system along the Mediterranean, into which the air is drawn. The bora consists of cold air blowing out from an anticyclone over Europe, and is very similar to the mistral of the Rhône Valley.

boulder clay see **till**

bourne A stream, especially on chalklands in southern England. Many bournes dry up for part of the year, normally in summer, when the water table falls.

braided stream A stream or river which splits into two or more smaller channels that rejoin further downsteam. The split or braiding is generally caused by deposits of sand or mud which have been dumped by the river. At times of flood the deposits may build up above the

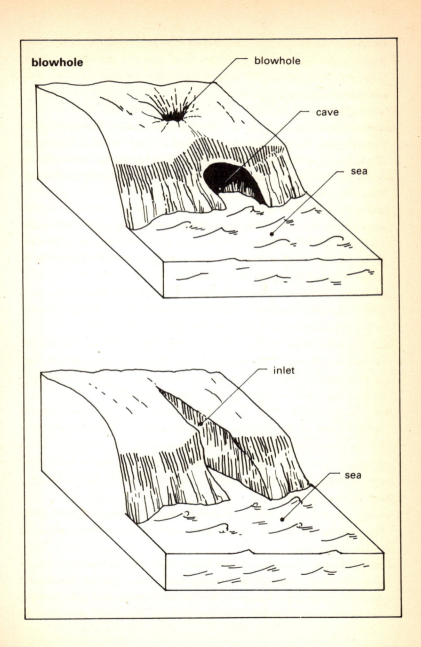

blowhole

blowhole

cave

sea

inlet

sea

29

normal river level, and then vegetation may grow on them. When this happens the small muddy islands become permanent. Braiding also occurs where streams are flowing out from melting ice and there are vast deposits of sand and gravel, through which the river finds several small channels.

Brandt Commission (1977) An international commission chaired by Willy Brandt, the former chancellor of West Germany, which studied the problems facing the Third World. The commission introduced the terms 'North' and 'South' to refer collectively to the more advanced countries and the less developed countries and suggested ways in which improvements could be made.

break of bulk The unloading and breaking down of a ship's bulk cargo for transportation by rail or road. Many ports have become break of bulk locations, which have also attracted industrial development because of the availability of raw materials. The term break of bulk is now applied to any change from one form of transport to another.

breccia A type of rock consisting of angular fragments which have been joined together by a fine-grained cementing material such as clay. The rock fragments are often of desert origin, but some come from volcanic eruptions.

Breckland An area of sandy heathland on the Norfolk–Suffolk border in East Anglia. Laid down by an ice sheet, it is an expanse of sandy infertile till, which is quite different from most of East Anglia, where the glacial deposits have formed many areas of rich soil. Much of Breckland has now been planted by the Forestry Commission, but some has been utilised as farmland, thanks to the addition of fertilizers.

brickearth see **loess**

bridging point A point where a bridge has been built across a river. It may be where the river narrows, where there is firm ground alongside the river for good foundations, or where there is an island in the middle of the river; the presence of an island always makes bridge building a little easier, as, for example, in Paris and Montreal. The lowest bridging point is the last bridge downstream. As this is also the first bridge upstream, it will possibly be the highest point to which boats can travel, unless the bridge can open to allow boats to pass through; for example, Tower Bridge, in London.

broad A small shallow lake in East Anglia, formed as a result of peat digging in the Middle Ages. Since the 1960s the Norfolk Broads have had problems with silting up and pollution, partly because of their increasing popularity as a tourist resort. In 1989 they were officially designated as a National Park, and efforts are being made to preserve the unspoilt areas and to restore the parts of the landscape which have been damaged.

broadleaf evergreen tree Any broadleaf tree, as opposed to a conifer, which retains green leaves all the year round. Most broadleaf trees, e.g. lime and ash, are deciduous and shed their leaves for the winter. There are some broadleaf trees which remain evergreen because of heat and moisture all year round, as in the Amazon Basin. A more temperate example of this can be seen along the southern coastal regions of the Atlantic states of the United States.

brown coal A type of coal, only 50% of which is carbon. Of poor quality, it burns with much smoke and leaves a lot of ash, and so is often converted into electricity at the mining location to save transport costs. Brown coal is an important source of energy in parts of the Soviet Union, Poland and East Germany. There is also a large mining area in southeast Australia. Also called **lignite**. See also **anthracite, bituminous coal.**

Burgess model* A concept of urban layout as defined by the American, E.W. Burgess. In the 1920s he studied Chicago and realized that cities often grow in a series of concentric rings. His five major zones were: 1 the central business district, 2 the transition zone, where factories were located, 3 a zone of cheaper housing where the factory workers would live, 4 residential homes which were more expensive, 5 the commuter zone, where the most expensive houses could be found, and the inhabitants could afford to travel into the city each day. His rather simple model has been criticized by many authorities, with justification, but it is surprising how many cities have some concentricity in their land-use zones.

bush In Australia, New Zealand, the United States and parts of southern Africa, an area of scrub where few people live and the landscape is still wild and isolated.

bush fallow see **shifting cultivation**

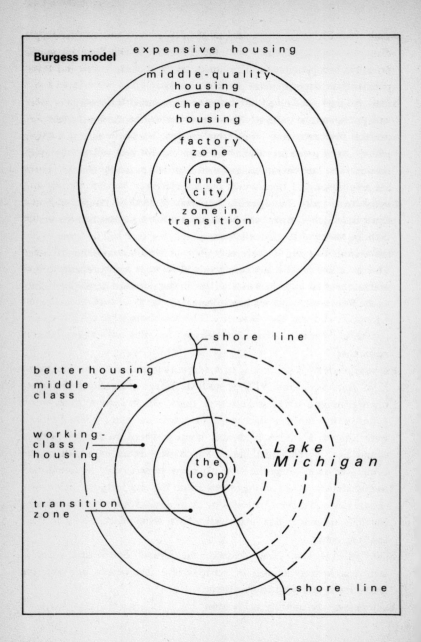

Burgess model

expensive housing

middle-quality housing

cheaper housing

factory zone

inner city

zone in transition

shore line

better housing

middle class

working-class housing

the loop

Lake Michigan

transition zone

shore line

bustee An area of shanty development in Calcutta. Thousands of people sleep on the pavements, but hundreds of thousands live in simple primitive accommodation. Huts made of cardboard, leaves and branches or even large drainage pipes are used.

butte A small flat-topped hill which is surrounded by deeply cut river valleys. A stratum of hard rock forms the top of the butte, and this protects the underlying layers from erosion. Formerly part of a larger plateau, it is a feature which has been cut off and isolated by river erosion. The best examples can be seen in the southwest United States. See also **mesa**.

Buys Ballot's Law The principle, observed in 1857 by Danish meteorologist Buys Ballot, that if an observer in the northern hemisphere stands with his back to the wind, the atmospheric pressure will be lower to his left than to his right, the reverse being true in the southern hemisphere. This is caused by the fact that winds blow from high pressure to low pressure and are deflected to the right in the northern hemisphere, but to the left in the southern hemisphere.

C

caatinga The thorny scrubland found in the interior of northeast Brazil. The vegetation, which includes cacti and acacias, is drought-resistant (xerophytic) because it has to be able to withstand prolonged dry spells, especially in the winter months.

cacao A small evergreen tree, *Theobroma cacao*, grown for its seeds, which are roasted and ground to produce cocoa. Found in equatorial areas, it needs a constant temperature of about 25°C and rainfall throughout the year; strong winds can be harmful. Pods, which contain the cocoa beans, grow out of the trunk and main branches. Among the world's principal cocoa-producers are Brazil, Ecuador, Ghana and Nigeria.

Cainozoic Also **Cenozoic, Kainozoic**. The most recent of the geological eras; the name means the recent period of life. It dates from about 70 million years ago and is the time when mammals become an important group. The era contains the geological periods of the Palaeocene, Eocene, Oligocene, Miocene and Pliocene, which make up the Tertiary; and the Quaternary, which is the Ice Age together with the 11,000 years since the last ice advance. During the Cainozoic there was considerable tectonic activity, which formed most of the present high mountain ranges, including the Himalayas, Alps, Andes, Rockies, Atlas and Caucasus.

calcareous The term used to denote a soil or rock containing calcium carbonate ($CaCO_3$). Chalk and limestone are calcareous rocks, though limestone also contains many other minerals. Soils which develop on such rocks are calcareous. See also **rendzina**.

calcite The mineral which is the main component of calcareous rocks; calcium carbonate. See also **calcareous**.

caldera A large crater formed by the collapse of the cone of a volcano. Subsidiary cones may form within a caldera as a result of subsequent eruptions. Crater Lake in Oregon in the United States is a large lake which has formed in a caldera.

Caledonian folding A major period of earth movements to form fold mountains, which occurred near the end of the Silurian period 400 million years ago. During this episode the Scottish Highlands and the mountains of Norway and Sweden were formed.

calorie The amount of heat energy required to raise the temperature of 1000 g water from 15° to 16°C. This energy can be generated by food and can be related to the amount of food required to keep a healthy human being fit and active. Starch, fat and sugar generally give more calories than lean meat, fruit and vegetables. It is thought that about 2500 calories would be adequate for each day, and 2000 would be the absolute minimum. An energetic person or a manual worker might require 3000 or more. In Britain the average daily intake is about 3200 calories and in the United States it is about 3500. Probably 60% to 70% of the world's inhabitants obtain 2000 or less, on an average day, and are suffering from hunger or malnutrition.

Cambrian A geological period which began about 600 million years ago and ended about 490 million years ago. It is the first division of the Palaeozoic era, and many rocks in North Wales are from this period. Similar rocks are found in northeast United States, which at that time was part of the same land mass as Wales.

canyon A deep river valley with steep sides. Downward erosion by the river has not been accompanied by rainwash erosion of the valley sides, since the harder layers of horizontally bedded strata protect the softer layers. Large canyons are found in the southwest United States, notably the Grand Canyon, which is nearly 2 km deep in places.

C.A.P. see **Common Agricultural Policy**

capital intensive The term used to denote an industry or business activity in which a relatively large proportion of money is invested in equipment or machinery, as opposed to labour. For example, highly automated car factories or cereal farming using combine harvesters are very capital intensive. See also **labour intensive**.

cap rock A layer of rock on the top of a plateau which is hard and resistant to erosion. The term may also apply to a hard stratum at the top of a waterfall, as for example at Niagara, or on the top of hills, as for example on the North Downs in southeast England where the clay is topped with flints.

carbonation The process by which rainwater containing carbon dioxide from the atmosphere causes chemical weathering of calcareous rocks by dissolving the calcium carbonate. The dissolved materials are carried away in the water, but may be deposited elsewhere to form stalagmites or stalactites.

Carboniferous A geological period, part of the Palaeozoic era. During the Carboniferous there were changing conditions, which gave rise to different types of deposits. At first much of Britain was covered by a clear warm sea containing many animals. When the animals died and fell to the seabed, they formed limestone; the carboniferous limestone found in the Pennines, Mendips and south Wales was formed at this time. Later the water became shallower, and deposits of sands accumulated, which formed the millstone grit rocks that can be seen in many parts of the Pennines. Later still, as the water became shallower, huge deltas were formed, on which there were forests of tree ferns. As the trees died, they were covered by mud and sand, so that instead of rotting away the trapped layers of dead vegetation gradually turned into coal. The three parts of the Carboniferous period; the limestone, the sandstone, and the coal measures, indicate the conditions that prevailed 350 to 270 million years ago, when these rocks were being deposited on the seabed.

cash crop A crop which is grown for sale to earn money, rather than for consumption by the farmer and his family. In the more advanced countries it is likely that the farmers will sell everything they produce, as for example the cereal growers on the prairies of Canada and the United States. Compare **subsistence crop**.

cataract see **rapids**

catch crop A crop which is grown just after a main crop has been harvested. It is likely to be a crop which can be grown and harvested in a relatively short time. The cultivation of catch crops is a way of making full use of the land; it is a source of extra income for the farmer and also ensures that the soil is not left exposed.

catchment area 1 The area which is drained by a river and its tributaries; the area from which a river catches or collects its water. A catchment area can be very small – just a few square metres – but it may cover an enormous area, as for example the Amazon Basin. Apart from some loss by evaporation, all the water that falls as rain within a catchment area will either run off on the surface or sink into the ground, eventually reaching the river which drains the basin. The speed at which the water reaches the river or its tributaries will depend on the vegetation cover and the rock type. If there are settlements in the area, the effects of man-made drains will also be relevant. Also called **drainage basin, river basin**. See also **watershed**. 2 The area from which people may come to a central institution, such as a school or hospital. Large towns will have large catchment areas, whereas villages are likely to have only a small catchment area.

cave A hollow space resulting from the erosion of rock. Caves are common in limestone areas where water dissolves the rock. Enlarged joints and bedding planes provide lines of weakness where erosion can set in. In many limestone areas there are underground streams, which contribute to the process of erosion and the formation of long networks of caves, caverns and tunnels. There are examples of limestone caves near Cheddar in the Mendips, near Neath in south Wales and near Ingleborough in Yorkshire. Caves also form on coastlines, where the exposed rock of cliffs is attacked and eroded by the sea. Corrasion is caused by the presence of sand and pebbles and, as well as corrasion of the rocks, there is likely to be hydraulic action. Air often becomes trapped in fissures in the rock; whenever a wave breaks the air is compressed, and this creates a hammering effect which helps to weaken the rock.

Celsius A widely used temperature scale devised by the Swedish astronomer Anders Celsius (1701–44). It is based on a 0 to 100 scale; water freezes at 0° and boils at 100°. Also called **centigrade**.

Cenozoic see **Cainozoic**

centigrade see **Celsius**

central business district The central part of a town or city, which is likely to contain many offices and business headquarters, as well as shops, banks, and possibly council offices. Land values are usually very high, and only the larger and wealthier organisations are able to pay the high rents or rates. The main advantage of the central business district is its central location, which makes it the most accessible part of a town or city. The retail and business functions which take place there are all helped by this accessibility. The demand for land explains why high-rise buildings and skyscrapers are commonly found in C.B.D.s. There are sometimes underground car parks; and buildings extend below the surface, as well as above, in order to make the fullest possible use of the available space.

central place 1 A settlement which is central to its surrounding area, and which provides goods or services to that area. 2 A shop or group of shops, as for example in a suburban area, which represents a centre to local inhabitants, though not necessarily to the town as a whole.

central place theory The theory that in an isotropic area settlements grow up at very regular distances, in order to serve the people who live in the surrounding rural region. It is associated with Walter Christaller, who in 1933 wrote about the settlements in his home area of southern Germany. His studies showed that there were large numbers of very small settlements, or hamlets, which he called first order settlements. Every few miles slightly larger settlements – villages – would grow, and at even greater distances towns would develop. Each settlement would be surrounded by an area which it served – its catchment area. Circular catchment areas give overlaps as well as gaps, and so hexagons have been used instead to represent catchment areas. The hamlets would offer only a small range of services, but the villages (second order settlements) would provide more shops. The towns (third order settlements) would have many more services and facilities, as they would be serving a far larger catchment area. A map can be drawn to show the hexagonal catchment area and the settlement hierarchy.

cereal Any of the cultivated grasses grown for their edible grain, such as barley, rye, wheat, rice and oats. Different cereals are grown in different environments according to the varying climatic conditions required for their growth: rice needs high temperatures (25°C) and wet conditions during the early months of growth; barley and rye can both survive in areas with very cold winters, providing summer temperatures reach 15°C. Barley grows well with an annual rainfall of only 550 mm but oats can grow in far wetter areas, with as much as 1500 mm annually.

chalk A soft sedimentary rock, which is very light in colour because it consists of the skeletal remains of dead sea animals. It is almost pure calcium carbonate in places and is a very permeable rock, which means that water can pass through quite easily. Athough chalk is soft, because of its permeability it does not always wear away very quickly, and sometimes stands up in steep slopes and cliffs. There are many chalk cliffs on the south coast of England, for example at Dover and Eastbourne, and also in Yorkshire, at Flamborough Head. There is little surface water on chalklands, although there are many dry valleys. These are the relics of river systems which used to flow on the surface, either because the water-table was then much higher, or because of permafrost conditions during a glacial phase of the Ice Age, when the frozen sub-soil prevented the water from sinking beneath the surface.

chaparral (In the United States, especially California) a low dense scrub consisting of small evergreen bushes. It is similar to the maquis of the Mediterranean lands.

chemical weathering see **weathering**

chernozem (*Russian*) A dark brown or blackish deep rich and fertile soil which occurs in areas of temperate grassland. The top layers of the soil are rich in humus, lime and most plant nutrients, and excellent crops of wheat have been grown for years on chernozems. Because the climatic conditions are quite dry, there is an upward movement of minerals through the soil. There are large areas of chernozem in the Ukraine and neighbouring parts of Romania and Hungary, and in the prairies of the United States and Canada. Also called **black earth**.

chert A rock made up of silica, probably derived from fragments of dead sea creatures. Chert occurs in chalky areas but, while chalk consists of the remains of sea creatures whose shells and skeletons were of calcium, chert is made up of those animals whose structural remains were of silica, such as sponges.

china clay A white clay formed from the disintegration of granite.

Obtained from feldspar, which is one of the minerals that makes up granite, it is used in the manufacture of high-quality porcelain. It has been quarried in many places in southwest England, including Dartmoor and Bodmin Moor, but the main source in Britain is near St Austell in Cornwall. It is also used medicinally and in the manufacture of certain types of paper. Also called **kaolin**.

chinook A warm wind which blows down the eastern side of the Rocky Mountains. It contains air which has come from the western side of the Rockies. As the air rises over the mountains it is cooled quite slowly – at the wet adiabatic rate – and rain or snow falls on the mountains, so that the air becomes progressively drier. When it descends the eastern side of the Rockies it gains heat at the dry adiabatic rate, becoming a warm wind by the time it crosses the prairies. It is commonest in winter and spring, when depressions come inland from the Pacific. The warmth of the chinook is often sufficient to melt snow and help to thaw out the soil, enabling cereals to be planted in the spring. It has been known for the chinook to change the temperature by as much as 20°C in about 15 minutes.

chitamene see **shifting cultivation**

C horizon The parent material from which a soil has been created; the subsoil. See also **A horizon, B horizon**.

choropleth map* A map with shading to provide quantitative information about different areas or regions. Choropleth maps show the average information relating to a given area, usually an administrative district, because of the availability of information and data. It is normal to draw choropleths in black and white, with white indicating the lowest numbers and black the highest. Dots and cross-shading may also be used for different classes of information, and each area of shading will have a definite boundary line, not a transitional area. Also called **shading map**.

cirque (*French*) a corrie

cirrocumulus A high cloud consisting of small patches of ice crystals often stretched out in a linear form. This type of cloud is called a mackerel sky.

cirrostratus A thin sheet of cloud which consists of high-level accumulations of ice crystals.

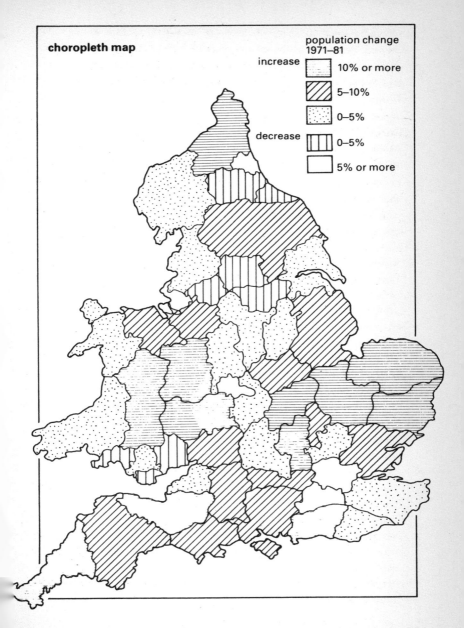

choropleth map

population change
1971–81

increase

	10% or more
	5–10%
	0–5%

decrease

| | 0–5% |
| | 5% or more |

cirrus A type of cloud which forms at great altitude, is thin, wispy or feathery in appearance and consists of minute crystals. Cirrus does not give rain, but its presence is sometimes a sign of an approaching depression. In this case the cirrus is gradually replaced by thicker and lower cloud. In different conditions, when patches of cirrus can be seen in a predominantly blue sky, it is a sign of good and settled weather. Cirrus cloud occurs at heights above 6000 m and sometimes as high as 12,000 m. See also **cumulus, stratus**.

citrus fruit The fruit of any plant belonging to the genus *Citrus*, such as oranges, lemons, limes and grapefruit. Citrus fruits require high temperatures for ripening, and so they are found only in warm climates, such as in the Mediterranean area or in tropical latitudes.

clay A sedimentary rock which consists of tiny particles less than 0.002 mm in diameter. The particles are very closely packed together, and so it is difficult for water to pass through; for this reason clay is impermeable. Clay soils quickly become waterlogged in wet weather. There are many different types of clay, varying in texture, chemical content and origin, such as alluvium, boulder clay and loess.

cliff A high steep, possibly vertical, rock face, commonly associated with coastal regions, but also occurring inland. The dip of the strata influences the steepness of the slope. If the dip is landwards, the slope is likely to be very steep but it will be more gentle if the cliffs are in hard rocks, such as granite, as at Land's End, or carboniferous limestone, as on the Pembroke coast in southwest Wales. Many other rocks form cliffs, as for example the chalk cliffs of Dover.

climate The overall weather conditions of a place or region throughout the year. The daily weather records are added together and then averaged out to give a general pattern of the climate. The climate of an equatorial forest is similar to the weather, but in most parts of the world the climate will be different from the daily weather. See also **continental climate, maritime climate, Mediterranean climate**.

clinometer An instrument for measuring inclination or slope.

clint A flat block of limestone between grykes. Clints are part of a limestone pavement and can be found in many areas where there are outcrops of carboniferous limestone, such as near Malham and Ingleborough in Yorkshire.

cloud A visible mass of water droplets or particles of ice suspended in the air above the earth's surface. See also **cirrus, cumulus, stratus**.

coal A sedimentary rock formed mainly in the carboniferous period, when dead tree ferns were buried beneath sediment in anaerobic conditions which slowed down the processes of decay. The gradual changes which occurred produced coal from the buried plant material. There are tree ferns growing in certain parts of the world today, such as Malaysia and New Zealand. If these deposits are covered by sediment, it is possible that they may form coal in a few million years. Coal is classified according to its hardness, carbon content, and the amount of smoke given off and the amount of ash left behind when burnt. See also **anthracite, bituminous coal, brown coal**.

coal measures The series of rocks from the Carboniferous period which contain seams of coal. Coal measures also contain many strata of clay and sandstone, and coal may account for only 2% of the total thickness.

cocoa see **cacao**

coffee Any tree or shrub belonging to the genus *Coffea*, grown for its seeds, which are roasted and ground to produce coffee. The coffee bush requires hot summers – about 25°C – for growing and ripening. It is an exhausting crop and requires a rich soil; in the major growing areas of Brazil many of the soils are from old weathered lavas. The main growing countries are Brazil and Colombia in South America, and Kenya in East Africa. Western European countries and North America are the chief importers.

cohort see under **age sex pyramid**

coking coal A high-quality bituminous coal which is converted into coke for use in blast furnaces in steelworks.

col 1 A gap or pass in a line of hills. 2 A calm area between two depressions and two anticyclones. It lasts only a few hours, as the depressions move in and push it away. It will cause sunny weather during the day, but fog or frosts at night.

cold front The dividing line between two air-masses, where a cold mass is advancing against and beneath a warmer mass. The upward movement of the warm air is accompanied by condensation, cloud formation and rainfall. At a cold front the temperature falls by 2° to 4°C, and the wind veers towards the northwest.

collective farming A system of farming found in the Soviet Union and other communist countries in which formerly separate smallholdings are worked as one farm, often thousands of acres in size. Each collective has a manager, who is responsible for organizing the work and purchasing machinery.

colony A settlement created by a group of people from a different country. For example, Britain established colonies in Australia, New Zealand and the United States. The French colonized parts of west Africa, and there were Portuguese colonies in southern Africa. Some of the colonies were created in uninhabited areas, but in other cases the colonists found that there were already native inhabitants in the areas of settlement, such as Aborigines in Australia, Maoris in New Zealand and Indians in North America. The colonial influence was not always beneficial; there were wars and bloodshed, and economic wealth was often taken from the colonies and shipped back to Europe. Administrators were not always helpful towards the native population. However, there were some good effects of colonization; for example, roads, schools and hospitals were built, and stable governments were usually created.

combe Also **coomb, coombe** A small armchair-shaped hollow found on a steep slope, generally an escarpment, in a chalk hillside. They are formed as a result of the opening up and enlarging of a joint by weathering. The weathered material slips to the bottom of the hill, where it may accumulate to form a mound of depositional material called **combe rock**. Much of the weathering is the result of solifluxion.

Comecon An economic and political union of several communist states. It was formed by the Soviet Union, Hungary, Bulgaria, Poland, Romania and Czechoslovakia, these countries being subsequently joined by East Germany, Mongolia, Cuba and Vietnam.

commercial agriculture The type of farming in which crops or animals are grown or reared for sale in order to earn money. Some commercial farms are quite small, but others are very large, highly mechanized and capital-intensive. See also **subsistence farming**.

Common Agricultural Policy (C.A.P.) The rules, regulations and guidelines which help farmers in the member countries of the European Economic Community. They are designed to enable farmers to earn a reasonable income and to encourage them to produce large quantities of essential foods. In several cases the C.A.P. has led to overproduction and consequent surpluses of some products, which cost millions of pounds to store. The main surpluses have been of wheat, milk, beef, butter, olives and wine.

common land Land which is normally used by commoners for grazing. Historically common land was not owned by any one person or family, but was available to members of the community. This is still true today of some village greens and riverside meadows, and in upland regions there are often areas of hillside which provide common grazing.

commuter A worker who travels daily from his home to his place of work in a large town or city. Efficient train services and good road networks are vital to enable large numbers of people to travel to and from work. Over one million people commute to London each day. A large proportion have journeys of from 30 to 40 mi, but some travel far greater distances – from Wiltshire, Oxfordshire, Suffolk and Northamptonshire, for example. A **commuter belt** is the area around a city in which the commuters live. It is generally an affluent area, which includes the outer suburbs and neighbouring small towns and villages.

compaction The compression of soil, as for example by walkers' feet along a popular footpath. When compaction occurs, it is difficult for rainwater to percolate through the soil, and so muddy patches and puddles develop. Compaction can also occur on farmland if heavy machinery goes over soft or wet ground.

component A constituent part of a product; one of the parts which have to be put together in order to make a finished product.

concave slope A slope which is much steeper at the top than at the bottom. On an ordnance survey map the contours are positioned much closer together at the top than at the bottom. When standing at the top of a concave slope it is possible to see the entire slope. Concave slopes are quite common on limestone hills. See also **convex slope**.

concealed coalfield A coalfield which is underground. The coal seams lie beneath strata of other rock types and have to be extracted by shaft mining. See also **exposed coalfield**.

concordant coast A coastline that runs parallel to the hills and structural features immediately inland. When submergence occurs the sea floods the former valleys and the peaks become chains of islands. Examples of a concordant coast are to be found along the south coast of Ireland, near Cork, in the province of Dalmatia, in Yugoslavia, and along the Pacific coasts of North and South America. Also called **Dalmatian coast, longitudinal coast, Pacific coast**.

condensation The process in which a gas or vapour changes to a denser form as a result of cooling. Mist, fog and cloud, for instance, are formed by the condensation of water vapour in the air. Dew is the result of water vapour condensing on grass or on the ground. Condensation forms tiny droplets of water, but they can merge and grow bigger and heavier; they may then fall as rain.

condensation nuclei Hygroscopic particles, such as dust or salt grains, which may attract water vapour and cause condensation before dew point is reached, thus producing mist, fog or cloud.

confluence The point where two or more streams converge and unite, forming a larger, probably faster-flowing, stream.

conglomerate A sedimentary rock consisting of large fragments which have been cemented together. The fragments are smoothed or rounded, and they are often mainly siliceous in content. Conglomerates form on beaches, where the smooth pebbles are an ideal source. See also **breccia**.

coniferous forest A forest which consists of cone-bearing, usually evergreen, trees with needles, such as pines (*Pinus* spp.), spruces (*Picea* spp.) and firs (*Abies* spp.). There may also be larch which, although coniferous, shed their needles and grow new ones each spring, and there are usually some silver birch in areas of coniferous woodland. Because conifers can withstand colder temperatures, coniferous woodlands are found further north than deciduous woodlands. The largest expanse is in northern Russia and Siberia in the Soviet Union, and further west in Finland, Sweden and Norway. Smaller areas occur on many mountainsides, for example, in the Himalayas, Alps and Rockies.

In Britain, the Forestry Commission has planted large areas of conifers. The major source of softwood for pulp and paper manufacture, conifers also provide timber for building purposes. See also **deciduous forest, evergreen trees, taiga**.

consequent stream   A stream that flows in the direction of the original slope of the land, that is, it follows the dip of the rock strata. Several consequent streams may develop more or less parallel to one another, each flowing down the same major slope. See also **subsequent stream**.

conservation   The careful and thoughtful use of resources. The natural environment should not be used in a wasteful way, and it is important to preserve and manage the landscape and the scenery, as well as the soil, minerals and wildlife. Planning and organization are necessary in order to make the best possible use of the countryside. The conservation of minerals involves using them more slowly, trying to reduce or eliminate waste and recycling anything which can be reused.

consumer industry   An industry producing ready goods for the consumer, such as food, clothing and domestic appliances, as opposed to goods which will be used in further manufacturing processes. Many consumer industries are located on trading estates. They use road transport and electricity, and this helps to make location very flexible. Southeast England is particularly important for consumer industries, because the largest markets are also in the southeast.

container port   A port for handling containers. Heavy lifting equipment and large areas of flat land are required for container ports. Container ships can be loaded or unloaded quickly because their entire cargo is boxed in containers, which are easily lifted straight on to (or off) trains and lorries. Examples of British container ports are London, Southampton and Felixstowe.

continental climate   The type of climate found in the interior of large continents, that is, hot summers with convectional rainfall and very cold dry winters, with only light falls of snow. Summer temperatures are about 20°C, and winter temperatures are −10° to −20°C in the coldest month. Total annual precipitation is about 500 mm. The natural vegetation consists of grassland. Areas of continental climate are found in Poland and Hungary, and on the steppes of the Soviet Union and the prairies of North America. In the southern hemisphere there are no areas of real continental climate because the land masses are relatively narrow, so that oceanic influences are present throughout. See also **maritime climate, Mediterranean climate**.

continental drift   The slow movement of the continents on the surface of the earth. The movement, or drift, occurs as a direct result of the movement of the underlying plates. Continental drift was first written about by Alfred Wegener in Germany in 1912. He noticed that the western parts of Africa, near the Gulf of Guinea, would fit alongside the coast of northeast Brazil, and discovered that there were geological similarities between the two areas. For a long time his ideas were discredited, but in the 1960s they were shown to be correct. It is now known that there used to be a vast ancient continent, which is referred to as **Gondwanaland**. About 100 million years ago, Gondwanaland fractured, and parts of it began to drift away to form the modern continents of Antarctica, Africa, Australia and South America. There has also been continental drift in the northern hemisphere, North America and Europe formerly having been joined together. See also **Pangaea**.

continental shelf   A shallow area of sea which is a gentle continuation of the neighbouring land. The depth of a shelf is less than 300 m; beyond that the water becomes much deeper at the continental slope, which descends to the deep ocean bed. A continental shelf may be quite narrow, as, for example, off the west-coast of South America, but in places can be more than 150 km wide; for example, in the North Sea and around Britain. The shallow waters contain rich food for fish, and therefore some of the world's major fishing grounds are on continental shelves, including those of Britain, Japan, and off the west coasts of Canada and the United States. Continental shelf areas, including the North Sea and Gulf of Mexico, have been exploited for oil and natural gas deposits. Unfortunately there are serious pollution problems in several of the shallow shelf areas of sea, notably the North Sea and Mediterranean.

contour   A line on a map which joins all places of equal height. On the 1:50,000 maps in the British ordnance survey series contours represent a vertical inverval of 10 m. Other maps may use a different interval, to suit the needs of the map. Coloured shading can be drawn on to maps, in the spaces between neighbouring contours; this technique is often used in atlas maps.

contour ploughing The practice of ploughing upland at right angles to the direction of the slope. Farmers use this method in order to conserve the soil, as horizontal furrows prevent or slow down run-off after rainfall. Sometimes different crops may be planted at different levels of the slope; in this way part of the land will still be covered by vegetation when some of the crops have been harvested. Contour ploughing helps to reduce sheet erosion as well as gullying. It is commonly practised in parts of central United States, such as Tennessee and Kentucky.

conurbation 1 A large urban area which has developed as a result of the expansion and merging of two or more urban areas. Urban sprawl can produce a continuous built-up area spreading for several kilometres. The official conurbations of England and Wales are London, Merseyside, Manchester, Tyneside, West Yorkshire, West Midlands and south Wales; Scotland has a conurbation round Glasgow. 2 A smaller, but continuous, built-up area.

convection rainfall Rainfall caused by convection currents. As the air rises it is cooled. When dew point is reached the water vapour condenses to form cloud. If the convection current is strong, a very high cloud may form, with great quantities of water and dust particles. The tall banks of cloud, **cumulonimbus**, can look very threatening; they will produce very heavy showers of rain, often more than 25 mm in the space of an hour or less. In tropical areas most of the rainfall is convectional; but in temperate latitudes convectional rainfall occurs only occasionally, in the summer after very hot anticyclonic spells of weather, and will be accompanied by thunder and lightning.

convenience goods Items which are quite cheap and are required frequently, possibly every day, such as bread and newspapers. Also called **low order goods**.

convex slope A slope which is steeper at the bottom than at the top. Standing at the top of a convex hill, it is impossible to see the bottom of the hill because there is always a large blind spot or dead ground.

Convex slopes commonly occur on chalk hills. See also **concave slope**.

coomb (coombe) see **combe**

cooperative An organization in which several small farmers pool their resources, equipment and marketing. In this way they have the purchasing and selling power of a large organization. A cooperative can help to provide more capital investment as well as a better and more secure income. Cooperatives were first developed in Denmark in the late 19th century, when many small dairy farmers began to work together in groups. Today they collect the milk from the farms, and either sell it or use it to produce cheese, butter or cream. Some of the skimmed milk is used for feeding pigs, and the bacon, lard, ham and sausages are also handled by the cooperatives. Many other countries, such as New Zealand, now have cooperatives, and in Britain the Milk Marketing Board works on similar lines.

coral A tiny marine animal which lives in massive colonies. When coral polyps die they leave a very small deposit of calcareous material, which forms limestone rock; this in turn may build up to form a coral reef. Corals which lived 300 million years ago can be seen as fossils in some of the Carboniferous limestone rocks of the Pennines. This indicates that at the time the rocks were being formed they were on the bed of a tropical sea, since most corals can live only in warm water. Some coral fossils have been found high on the slopes of Mt Everest.

core 1 The central part of a region. The outer or more isolated parts are called the **periphery**. John Friedman wrote about the **core-periphery idea** after his studies of economic development in Venezuela. The core is the area where most of the economic development and growth take place. In Third World countries it may be the only area with an infrastructure of roads, electricity, water supply, etc. In advanced countries all aspects of the infrastructure will be more widely available, but even so, there are particularly favoured areas where growth is more likely to occur. The core in England would be the southeast, around London. On a larger scale, the core in the European Community would be a zone encompassing southeast England, Belgium, the Netherlands, northeast France and western Germany. It is often difficult to create developments in peripheral regions because there will be transport problems, and possibly labour shortages. Governments often give aid to encourage companies to set up industries in the more isolated places. In Britain there have been Assisted Areas and Enterprise Zones, where the cost of setting up a factory has been artificially low. See also growth poles. 2 (In physical geography) the inner part of the earth, where temperatures are very high. The rocks are mostly made of iron and nickel.

Coriolis force A force resulting from the rotation of the earth which deflects moving bodies to the right in the northern hemisphere and to the left in the southern hemisphere. This affects winds, ocean currents and humans.

corn 1 (In Britain) any of the principal cereal crops, that is, wheat, oats or barley. 2 See **maize**.

corn belt see **maize**

corrasion Mechanical erosion caused by loose material, such as sand or pebbles during transportation. As this material is carried by wind, water or ice, it scrapes against the bed or sides of the river or glacial valley, or against the cliffs and the shore. Corrasion also occurs in deserts, where it is caused by wind-blown sand. The sand and pebbles carried in the load of a river can cause both vertical and lateral corrasion.

correlation A statistical technique for comparing two or more sets of data. The variables can be compared in a precise and quantitative way. There is generally one dependent variable, the others being independent. The strength of the link or relationship is quantified between +1 and −1. A strong inverse relationship would be −1, while +1 would indicate a strong direct relationship; 0 would reveal the absence of an explanatory relationship. See also **Spearman rank correlation**.

corrie* An armchair-shaped hollow on a hillside, which has been cut by ice and freeze-thaw activity. If the hollow is deep enough, a lake will form in it when all the ice has melted. The sides and back wall of a corrie are steep and may be several hundred metres in height. There are several corries in Snowdonia, the Lake District, and many parts of Scotland, and they occur in most high mountain regions, including the Alps, Andes and Rockies.

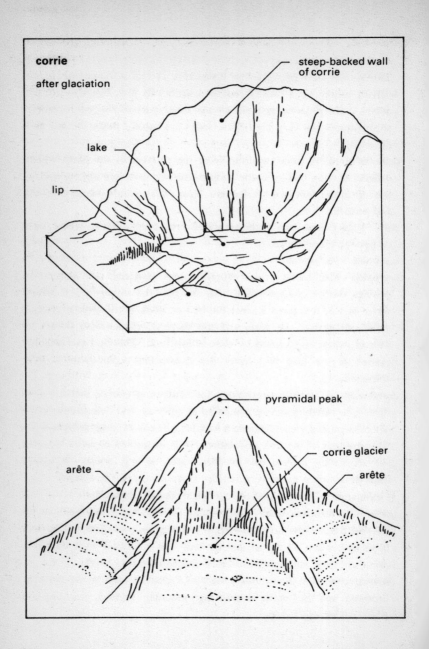

corrie

after glaciation

steep-backed wall
of corrie

lake

lip

pyramidal peak

corrie glacier

arête

arête

corrosion Chemical erosion leading to the disintegration of rocks, such as by the action of running water or by solution. Corrosion is particularly common on limestones.

cottage industries Small-scale industries which are carried on in the home of the worker. Many woollen and cotton goods, as well as baskets and pots, are made in this way. They are generally marketed and sold by a middleman or cooperative organization. Cottage industries used to be common in the Scottish Highlands and in Ireland, and there are still examples to be found. India has many small-scale domestic industries, and they can also be seen in many parts of Africa, as the rural people try to earn a little extra money.

cotton Any bush or tree of the genus *Gossypium* grown for the fluffy white fibrous substance which surrounds its seeds. The cotton bolls (seed heads) are still picked by hand in many places, since it has proved difficult to develop a machine which will pick only the ripe bolls, leaving the unripe heads to mature. After picking, the cotton is ginned to clean it. Then it has to be combed and spun into thread before it is finally woven into cloth. The major cotton-growing area is the cotton belt of the southern United States, comprising Alabama, Louisiana and Texas. The subtropical climate of this area means that summer temperatures are high (25°C), and the growing season is long, with over 200 frost-free days. There is a normally dry period before harvesting is due, which is useful, as too much rain can damage the boll. Much of the cotton is grown on plantations which used to employ slave labour in the 19th century. There are still large estates in the cotton belt, but they also grow other crops, such as corn, soya bean and peanuts, and often raise cattle.

counter urbanization The movement of people away from cities into smaller towns and villages. Counter-urbanization is a trend among the wealthier sections of the population in North American and in western European countries and is not associated with Third World countries, where there is still a strong flow of people into the cities.

country code A code of practice designed to protect the rural environment:
Guard against all risk of fire.
Fasten all gates.

Keep dogs under proper control.

Keep to the paths across farmland.

Avoid damaging fences, hedges, and walls.

Leave no litter.

Safeguard water supplies.

Protect wild life, wild plants, and trees.

Go carefully on country roads.

Respect the life of the countryside.

Countryside Commission An organization established in 1968 to replace the National Parks Commission. It has been an independent body since 1982, funded by the Department of the Environment. It is responsible for the conservation of areas of natural beauty in England and Wales, and aims to encourage the provision and improvement of facilities for enjoyment of the countryside and access for open air recreation.

cover crop A crop grown to protect the soil from heavy rain, wind, and erosion. Often consisting of a quick-growing plant, it may be grown after the main harvest of cereals, when large areas of land would be left bare. Sometimes strips of land on a hillside are left with a cover crop, whilst the remainder of the slope is being harvested. The most useful cover crops are legumes because they grow quickly, have good rooting systems for holding the soil, and they also add nitrogen to the soil.

crag A steep-edged rocky outcrop on a hillside. It will be the result of erosion and weathering, and is most likely to be found in areas of hard rock. Glaciated areas often have exposed rocks on truncated spurs.

crag-and-tail A crag with a tail of glacial material on the down-ice side. The debris may have accumulated as a glacier or an icesheet scraped over the hard lump of rock. Alternatively, erosion may have occurred on one side only, the down-ice side having been protected by its situation. An outstanding example is the crag-and-tail on which Edinburgh castle is located.

crater 1 A depression at the top of a volcano. Whether at the summit of a mountain or simply as a hollow in the ground at low level, it will form the top of a tube or vent through which the erupted material passes to the surface. When the volcano ceases to erupt, the vent and the crater may be filled with solidifying rocks. Examples of low-level craters can be seen in the Eifel district of West Germany. 2 A hollow on the earth's surface caused by a falling meteorite. There are also many craters on the moon.

crater lake A lake which has formed in a volcanic crater. If the volcano is still active, any eruption will begin with the ejection of large quantities of water out of the crater. Crater Lake in Oregon is an outstanding example of a crater lake. Other examples can be found in Indonesia, Italy and Germany.

creep The slow downhill movement of soil or regolith. The movement is caused by gravity, but is helped by water acting as a lubricant. The rate of movement is very slow, measured only in millimetres per month, or even per year. The soil which has slid downhill may accumulate against walls or other barriers and at the bottom of steep slopes.

crevasse A vertical crack or fissure in ice. It may be several metres in depth and will continually change as the ice moves downhill. Crevasses form when the slope of the bedrock over which the ice is moving becomes steeper. If the gradient lessens then the crevasses tend to close up. Crevasses caused by a change of slope normally run across the valley and are transverse. Longitudinal crevasses form when the valley broadens out and the ice stretches to fill up the extra space. As crevasses move downhill, they change angle slightly, and so it is possible to have crevasses running in any direction. Crevasses are largest on mobile glaciers and are potentially very dangerous for anyone crossing a glacier or icesheet. See also **icefall**.

croft A small farm in northwest Scotland, on the Scottish islands, or in western Ireland. It consists usually of a house with a small plot of land, where potatoes, hay, and oats may be grown, and there will be grazing land for sheep and possibly a few cattle. The crofters will grow most of their own food, and possibly make their own clothing, from wool. They will probably get additional food from fishing. Nowadays, many crofters are now involved in the tourist industry, and so they can earn some money; but fifty years ago they were subsistence farmers.

cross-section A technique used in map work to show relief in a clear and simple fashion. Cross-sections taken from ordnance survey maps have a horizontal scale equivalent to that of the map, namely 1:50,000 or 1:25,000. The vertical scale is exaggerated in order to show up the undulations. However, the exaggeration must not be too great; otherwise quite small hills will look like the Himalayas. A suitable vertical scale would be 1 cm to 100 or 200 m.

crude birth rate see **birth rate**

crude death rate see **death rate**

crust see **lithosphere**

cuesta (*Spanish*) A hill with a steep slope (**scarp**) on one side, and a more gentle slope (**dip slope**) on the other. A cuesta has an asymmetric cross section. It may have been formed by folding, but is more likely to be the result of tilting, followed by erosion, different rocks having eroded at different rates to form the steep scarp on the exposed edge of a resistant stratum of rock. Examples can be seen in the North Downs, South Downs, Chilterns and Cotswolds.

cumulative causation The theory that an initial advantage can produce a sequence of developments, which in turn lead to further advantages and developments. Cumulative development of this kind occurs in many industrial regions. For example, the coal which gives rise to a steel industry may also support a chemical factory, which would also help steel-making. Then engineering works may be set up to provide equipment for the steel and chemical industries. Meanwhile electricity supplies will have been laid on, and schools and hospitals, and roads and railways will have been built. Anyone then wishing to start a new enterprise would choose to go to this area because of the amenities and advantages it has to offer. Gunner Myrdal, who formulated the theory of cumulative causation, devised a development model following this principle. In Third World countries most developments occur near the major city, or cities, which are the only locations with infrastructures. The accumulation of development in the favoured areas leads to great regional inequalities in all countries. An attempt to develop new areas requires new growth poles.

cumulative frequency curve A frequency curve produced by adding the value of each successive group to that of the last. For example, if 15% of the inhabitants of an area were aged under 15, another 15% were aged 15–25, and a further 10% aged 25–30, etc., the points plotted on a graph would be 15, 30 (15 + 15), and 40 (15 + 15 + 10).

cumulonimbus see **convection rainfall**

cumulostratus see **stratocumulus**

cumulus A dome-shaped usually flat-bottomed cloud of considerable vertical growth. Small white cumulus are associated with fair, dry weather, but large towering cumulus may develop into cumulonimbus, which bring heavy rainstorms.

cut-off see **oxbow**

cwm A Welsh name for a corrie.

cycle of erosion The continuous process by which landscapes are formed and then gradually transformed by erosion. Mountains are formed by uplift, folding, faulting and volcanic activity, all of which are related to plate tectonics. Once the high land has been formed rivers, ice and all the agents of erosion and weathering begin to denude the landscape. When it is still high, but dissected by steep V-shaped valleys, it is described as a young or youthful landscape. Various methods of transportation remove the eroded materials and expose more rocks for denudation. Gradually the high land will become lower and the valleys will widen to form a mature or middle aged landscape. Eventually, after millions of years, most of the high ground will have been eroded and the river valleys will be broad plains. This old-age landscape, with mostly flat and low ground but just a few isolated residual hills, is called a peneplain. The sequence of events described is a very simplified and theoretical view of landscape evolution, because uplift and erosion take place simultaneously.

cycle of poverty* A chain of circumstances in which poor people who are unable to grow much food have little energy for work and therefore cannot earn money or do more farming in order to produce the food they need. They struggle for survival from one year to the next, are trapped in a vicious circle and cannot escape without help from somebody else.

cyclone A system of atmospheric low pressure which occurs when a cold air mass moving southwards from the Arctic meets a warm air mass moving northwards from the tropics to form a circulating air mass characterized by relatively low pressure at the centre, and by anticlockwise wind movement in the northern hemisphere and clockwise motion

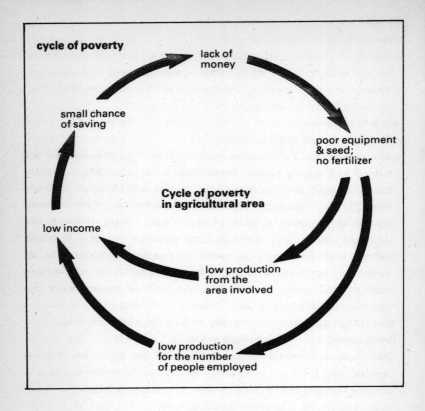

cycle of poverty

lack of
money

small chance
of saving

poor equipment
& seed;
no fertilizer

**Cycle of poverty
in agricultural area**

low income

low production
from the
area involved

low production
for the number
of people employed

in the southern hemisphere. There are two types of cyclone: the depression, which is associated with temperate latitudes; and the tropical cyclone, which is much more violent, but usually affects a smaller area.

cyclonic rainfall Rainfall associated with an area of low atmospheric pressure.

D

dairy farming A system of farming in which cattle are kept for their milk, which may also be used for butter, cheese, cream and yoghurt. It is normally intensive and dairy farms are relatively small. Rich grass is needed to feed the cattle and mild winters are an advantage, so that the cattle can stay outside all year. In southwestern England the mild winters caused by the North Atlantic Drift enable grass to keep growing and reduce the amount of feed which has to be bought. In many parts of England dairy cattle have to be kept indoors and fed for three or four months each year, and in some mountainous areas such as the Alps, the cattle are kept indoors for up to six months. For this reason many farmers try to make as much silage and hay as possible during the summer months, to give to the cattle in winter.

dale A valley in northern England or southern Scotland.

Dalmatian coast see **concordant coast**

dam A barrage, especially one associated with the generation of hydro-electric power. Some dams may be enormous, for example, Aswan in Egypt, and create lakes over 50 km long, whereas others may be no more than a few metres in length and form a lake which dries up for half the year.

Date Line see **international date line**

death rate The number of deaths in a year for every 1000 head of the population. In Britain the figure is 12, in the United States it is 9, but in poorer countries the figure is often higher; Bangladesh, for example, has a death rate of 19. Also called **crude death rate**.

deciduous forest A woodland consisting of trees which shed their leaves. Most of the trees associated with deciduous woodland are broadleaved, such as oak, ash, beech, etc, and in temperate latitudes they shed their leaves annually in autumn because the winters are too cold and frosty for growth to continue. However, in some monsoon lands, such as India and Burma, there are trees which shed their leaves in the hot season in order to avoid the excessive loss of moisture caused by transpiration. Near the equator, because of the constant climatic conditions, the deciduous trees lose their leaves sequentially rather than in one season, since there is heat and moisture all the year round. Although the trees shed some leaves each month, they are able to grow new ones too, as if the trees were going through all four seasons at the same time. See also **coniferous forest, evergreen trees, taiga**.

deflation The removal of tiny particles of dust or sand by the wind. It is a common process in deserts; in places deflation removes all the depositional material, leaving a surface of bare rock. Deflation hollows can be formed in areas of sand-dunes in coastal regions when the wind transports large quantities of sand.

deforestation The cutting down and removal of trees, usually to clear the land for some other use, such as agriculture, urban construction, industry, or road-building. Much of Britain was cleared for agriculture during the Middle Ages. Nowadays deforestation is occurring mostly in Third World countries, as new land is being opened up for agriculture. In many places deforestation has exposed the soil to heavy rain, which has led to severe erosion; as, for example, in many parts of southern Africa, northeast Brazil and the Amazon basin. Where steep slopes have been deforested, soil erosion has become a problem in many countries, including Italy and New Zealand. In some fairly dry regions the soil erosion has been particularly serious, since desertification has followed; for example, in the Sahel. Another result of deforestation is increased flooding. Trees and their roots slow down the rate of run-off and through-flow, but in places which have lost their trees run-off is much faster and floods are more extensive. This is true in many parts of southeast Asia and also in Amazonas province in Brazil. The removal of forests on Himalayan slopes in Nepal and northern India has caused damaging floods in the Ganges valley in India, and over much of Bangladesh.

delayed run-off The proportion of rainfall which sinks into the ground and gradually percolates through the rocks and back into a stream or river. See also **surface run-off**.

delta An accumulation of silt deposited on the seabed at the mouth of a river. When a river reaches the sea its speed is reduced, and so it has to deposit some of its load. If this deposition builds up above sea level it will form mud banks, which may grow into a delta. Deltaic deposition will fill in an estuary and gradually spread seawards. Vegetation will colonize the depositional area and convert the sediments into firm ground on which farming can take place. Deltas form most easily in shallow seas, and also where the tides and currents are not very strong; otherwise the depositional material may all be carried away. Deltas do not form around the shores of Britain, possibly because of the high tidal range, although there are some similar depositional areas where land has been reclaimed behind spits; for example Borth spit at Orfordness, and in some lakes in the Lake District and elsewhere there are deltas which were formed by rivers. Very large deltas covering hundreds of square miles are found along some coasts; as, for example, those of the Ganges, Mississippi and Nile. Deltas can be triangular in shape, such as the Nile delta, but others, including the Mississippi, are called bird's-foot deltas because they create narrow mounds of deposition protruding out to sea. Many deltas become important for farming as they have been created by alluvial deposits, which are generally a source of rich and fertile soils. The Ganges delta in Bangladesh is very productive, though often affected by flooding.

demersal fish Fish which live near the bottom of the sea, such as cod, halibut, plaice, sole and whiting; they are caught by trawling.

demographic transition The change in the pattern of population growth. In agricultural societies in a pre-industrial era there are high birth rates and high death rates, so the population remains fairly steady. Britain was at this stage before the Industrial Revolution, and some parts of Africa have not yet advanced beyond it. With improvements in education, diet, medicine and some technology, the high death rate begins to fall rapidly, but the high birth rate remains. This means that there is a rapid increase in population. Many African, Asian and Latin American countries have just gone through this stage of development. In the more advanced countries education continues to improve and a high proportion of the population now live in the urban areas.

Standards of living are higher, there is more employment and income and there is no longer the necessity to have large numbers of children. The birth rate falls and gradually becomes closer to the death rate. Eventually the birth rate will be comparable to the death rate and the population stable. See also **zero population growth**.

dendritic drainage A drainage pattern which looks like a tree when viewed from above. The main river is like the trunk; the major tributaries represent the branches; and the minor tributaries are like the twigs. Dendritic drainage patterns are most likely to develop in areas where the rock types are similar so that the rate of erosion is fairly uniform. An example of dendritic drainage is the Mississippi.

denudation The wearing away of land by weathering and erosion. Denudation is a broad term and includes all the natural agencies, such as sun, rain, wind, rivers, frost, ice and sea as well as heating and cooling, freezing and thawing, solution, abrasion, corrasion and corrosion. In addition to weathering and erosion, the removal of material, that is, the transportation, is also part of denudation. Together with deposition, denudation is the major process which creates the earth's landscape.

deposition The laying down of material which has been removed by denudation. Most of the material will be sediment and therefore a possible source of sedimentary rocks when consolidated. After denudation has taken place, the material is transported, sometimes over considerable distances, before being deposited. Most of the material will eventually be dumped on the seabed by rivers. However, there will be some deposition on land, for example, silt on flood plains, till deposited by ice, loess deposited by wind. Together with denudation, deposition is the major process which creates the earth's landscape.

depressed region A formerly thriving and successful industrial area which has declined as a result of one or more of its industries becoming uneconomic. Several of the old coal and steel regions of Britain, including south Wales, northeast England, and central Scotland, are examples of depressed regions, where the decline of industry led to high rates of unemployment. See also **development area**.

depression An area of low atmospheric pressure in temperate latitudes;

a cyclone. Depressions are formed when two contrasting air masses come into contact with one another to form a circulating mass of air. In the centre the air rises, and the atmospheric pressure is lowest, usually 970 to 980 mb, where the air is rising most strongly. The two air masses which form the depression will be associated with two fronts – a warm front and a cold front. The warm front is the leading front and gives a belt of cloud, and possibly rain. The temperature rises at the front and the wind direction changes from south-easterly to south-westerly. These marked changes of wind direction and temperature are called **discontinuities**. Behind the warm front is the warm sector, where a low layer of stratus cloud will give grey and gloomy weather, but probably not any rain. This is followed by the cold front, with tall cumulonimbus clouds and the likelihood of heavy rain. Behind the cold front will be brighter, showery weather, but colder and accompanied by north-westerly winds. Gradually the cold front catches up with the warm front to form an occlusion. Depressions move at 30 to 50 km per hour and generally travel from west to east or southwest to northeast. They occur in all the major oceans in temperate latitudes and generally cover an area of 800 to 3000 km. Deep depressions with lower atmospheric pressure can give rise to winds of 80 to 160 km per hour, but shallow depressions are associated with winds of 30 to 50 km per hour.

desalination The process of removing salt from sea water. In a number of areas, where there is a shortage of fresh water, salt is extracted from salt water by means of desalination plants. They also produce piles of salt in addition to the fresh water. Desalination plants can be found on islands such as Lanzarote and in some desert countries such as Kuwait and Saudi Arabia.

desert An area with a dry climate, sometimes defined as having a total annual rainfall of 250 mm or less. The effectiveness of 250 mm can vary, however, depending on whether it falls within a short or over a prolonged period. A desert is almost barren, although there are very few places with absolutely no vegetation; this occurs only on some patches of moving sand-dunes, or on bare rocky areas. Most deserts contain tufts of grass and scattered, usually thorny, bushes which have the ability to withstand long dry spells. In some deserts there are cactus plants, and various species of flowering plant which have a very short life cycle. They spring up quickly after a shower of rain, and go through their complete life cycle in a few days, before dying down and leaving seeds to lie dormant until the next rain. Some deserts contain large areas of sand-dunes, but most deserts are rocky. There are bare rock areas, but generally there will be a cover of loose stones with patches of moving sand. The stones and sand are gradually broken down by mechanical weathering and wind action. The world's major deserts including the Australian desert, Kalahari and Sahara, are found in the horse latitudes, where the permanent high pressure causes drought throughout the year. Deserts occur in various west coast areas where the influence of cool currents offshore makes the land even drier, with less than 100 mm rainfall in places, for example, Atacama in Peru (Humboldt current offshore), and the Namibian desert in southern Africa (Benguela current). There are deserts in the middle of the largest continents where no onshore winds can reach to bring any rainfall, for example, Gobi in Asia, and smaller examples in Nevada and New Mexico in the United States. Some cold regions are also regarded as deserts because they have less than 250 mm annual precipitation. Antarctica and Greenland are ice deserts, and even the tundra areas of northern Canada, northern Siberia, and elsewhere have some similarities to deserts, although they have cool summers and very cold winters.

desertification The process by which a desert gradually spreads into neighbouring areas of semi-desert, transforming them into true desert. The change may result from a natural event, such as the destruction of the vegetation by fire or a slight climatic change, but it occurs most frequently as a result of human activity. In many semi-arid areas and dry grasslands the vegetation becomes overgrazed by domestic animals, so that the land is left bare. Wind and rain then erode the soil, removing any residual fertility, so that no new vegetation can survive. Once the vegetation and soil have gone, the land becomes desert. An additional problem is the use of trees and bushes for firewood by the local people. They must have wood to make the fires they need for cooking. With an increasing population there is an increasing need for fuel, and so landscapes become depleted of trees. Once it has been destroyed, restoring the land is a long and slow process. Extensive desertification has been taking place in the Sahel, the southern edge of the Sahara, and there are also examples in the Soviet Union, southern Africa, India and many other places where there are too many people trying to eke a living from an inadequate landscape. The land needs to be rested and the size of herds restricted, if more land is not to turn into desert.

destructive wave A wave action which has a destructive effect on the shoreline because the backwash is more powerful than the swash. If a series of waves comes up a shore in quick succession, they break with an almost vertical plunge, washing shingle and sand downshore, and removing material out into slightly deeper water.

developed countries Those countries with a highly developed industrial sector and a high proportion of the population living in urban areas. Agriculture is likely to be commercial and mechanized, with subsistence farming constituting a minority activity. Infrastructure will be widely spread. Most of Europe, North America, Australasia and Japan could be described as developed, as well as parts of China, Africa and Latin America.

developing countries The poorer Third World countries in which numerous economic developments continue to take place. Brazil, Nigeria, and many other countries have been developing rapidly, as industrial growth and urban growth have shown some similarities to western Europe and North America. Loans from the World Bank and foreign countries are often vital for such developments, and several of the developing countries are now heavily in debt.

development area (In Britain) a depressed area of high unemployment where the government provided finance and subsidies to encourage the development of new industries. The main areas concerned were central Scotland, south Wales, the west Midlands, parts of Lancashire, and parts of Yorkshire and northeast England. New towns were created and many trading estates were built. Development areas no longer exist in their original form, but many of them are now called assisted areas and can receive government help with a variety of development schemes.

dew Moisture condensed from the air and deposited on grass and other plants, especially during the night. Dew is formed in calm and settled conditions generally associated with anticyclonic weather. During the night, the temperature falls and the air is unable to hold as much water vapour; some condenses on to plants as dew, and some may form mist or fog. The cooling which causes dew is the result of radiation during the night. The air cools from the ground upwards, and if dewpoint is reached condensation will occur. It is most frequent after sunny days, especially in autumn. Dews are very common, and heavy, in deserts, and there are some plants which survive on the moisture from dew. If the temperature falls below freezing point, the dew would become hoar frost.

dewpoint see under **absolute humidity**

dew pond A small, usually artificial, pond in which water can accumulate for sheep and cattle. Normally lined with clay or concrete in order to hold water, they are found on the chalklands of southern England and on some limestone hills. In spite of the name, the water is derived mainly from rainfall, not dew.

differential erosion The type of erosion which occurs as a result of different types of rock wearing away at different rates. Coastlines of erosion are rarely straight because of differential erosion, and it is a phenomenon which also affects inland landscapes. Hills and undulations are the result of variations in resistance to erosion shown by different types of rock. Igneous intrusions are normally quite hard, and so sills and dykes generally form higher ground than their surroundings. However, in western Scotland, for example, there are some igneous intrusions which are softer than the adjacent rocks and they form hollows.

dike see **dyke**

dip The steepest angle of a tilted stratum of rock. It is measured from the horizontal, 90° being vertical.

dip slope see **cuesta**

discharge The water flowing down a river channel. The discharge is expressed as Q = AV.

Q is the discharge, generally in cubic metres per second.

A is the cross-sectional area of the channel.

V is the mean velocity of the stream.

The mean discharge of the River Amazon is 170,000 m^3s^{-1}.

discordant coastline A coastline that runs at right angles to the structural features of the landscape immediately inland. It shows up most clearly where there are lines of hills or mountains running inland. If submergence takes place, the valleys will form large inlets, known as fjords or rias, depending on how they were originally formed. There are several good examples of discordant coasts on the edges of the Atlantic Ocean, for example, in southwest Ireland, Brittany and northwest Spain. Also called **Atlantic coast, transverse coast**.

dissected plateau see under **plateau**

distance decay The pattern of diminishing influence of a town, city or the like in proportion to its distance from a neighbouring town, etc. This is because of transport problems and the time taken in travelling from one place to another. The attraction of a zoo, country park, etc. lessens with distance.

distributary A branch which flows from a main river and does not rejoin it, distributing the water and eventually carrying it to the sea. Distributaries may be narrow and shallow, but in some cases are large enough for navigation, as for example, in the Ganges delta.

diurnal range The difference between the maximum temperature during the day and the minimum temperature at night in a 24-hour period.

diversification The broadening out of a business or industry to encompass a greater range of goods or activities. In an industrial area where there has been a concentration on one speciality, it is advisable to develop different industries, in case the main industry suffers from a slump. This happened in many old coal and steel areas, such as south Wales, and so a variety of industries were set up on trading estates. In agriculture monoculture can be very successful commercially for a few years, but one pest or disease may destroy the entire crop. It also seems likely that repeatedly growing the same crop is harmful to the soil. By growing a variety of crops (and perhaps introducing a variety of animals) the danger from pests can be reduced, and by rotating crops the quality of the land can be maintained.

doldrums The low pressure area near the equator. The overhead sun keeps the equatorial latitudes permanently hot, and so there is always rising air to create low pressure. The rising air causes convection storms, and so rain is frequent, often daily. For this reason these latitudes are naturally covered with forest. Because of the seasonal changes in the position of the overhead sun, the main centre of low pressure moves a few degrees north of the equator in June and a few degrees to the south in January. The northeast and southeast trade winds converge in the doldrum belt, contributing to the mass of rising air. The meeting point of the two trade winds is the **intertropical convergence zone**.

dormitory town A settlement a few miles from a large city in which many of the inhabitants are commuters. There will often be a shortage of shops and amenities in the dormitory settlement. Dormitory settlements can be found around most large cities.

dot map A map on which dots representing a specific value are used to show distribution. If showing the distribution of sheep, for example, one dot may represent 50, or perhaps 100, sheep. The dots should all be the same size. When drawing a dot map it is useful to know something about the subject matter; for example, if trying to show sheep distribution in Cornwall and Devon, it would be useful to know that most of the sheep in the area are located on Dartmoor and Bodmin.

downland An area of hilly pasture, especially in Australia and New Zealand.

downs Open, rolling chalk uplands, especially in southern England. Mainly treeless grassland with thin soil, downs are used traditionally as permanent pasture for sheep, although some areas have been ploughed up to grow cereals, especially barley.

drainage The removal of water, whether by natural means, such as rivers, or by artificial means, including drainage pipes and channels. Natural drainage includes the throughflow and infiltration as well as the channelled flow in rivers.

drainage basin see **catchment area** (def. 1)

drift 1 Material which has been transported and then deposited by ice. It may consist of clay, sand, or larger gravels, or a combination of all these. It can also include fluvial glacial materials which have been transported by melt water. 2 A broad slow-moving ocean current, for example, the North Atlantic Drift. 3 See **continental drift**.

drift mine A type of mining in which a sloping tunnel is dug into the ground to give access to the mineral that is to be extracted. This method is employed at Selby coalfield in Yorkshire.

drizzle A light form of rain consisting of small droplets which are only just heavy enough to fall. It is generally associated with stratus cloud.

drought A prolonged period of dry weather. Officially, a drought in Britain means 15 consecutive days without measurable rain. There are, however, different definitions in different parts of the world. The degree of aridity of an area will depend on the temperature and the amount of evaporation, as well as on the amount of rainfall. Drought is experienced annually in some regions, such as parts of Ethiopia and the Sahel. Some regions of India may have sufficient rainfall for crop growing some years, and then may have a year with less rain than is needed to grow crops. This occasional drought can be more serious than the droughts which occur every year in some regions. The severity of a drought will vary from country to country. A drought in Britain may mean that restrictions will be placed on watering gardens or using hosepipes for washing cars, but a drought in some African countries may mean that there is starvation.

drowned coastline A coastal strip which has been submerged under the sea, either because the sea-level has risen or because the level of the land has sunk. Valleys become flooded, creating rias, and hills may become islands.

drowned valley A submerged valley, which has been flooded by the sea or a lake. Sometimes valleys are flooded when dams are built. Lakes created in this way may not only drown farmland but also drown villages, as for example, Ladybower reservoir in the Peak District, in which parts of a village occasionally become visible in very dry summers.

drumlin An elongated mound of glacial till. About 500 m in length and 10 to 20 m in height, a drumlin consists of unstratified glacial debris. Drumlins usually occur in groups, called **swarms**, which are sometimes described as basket-of-eggs topography because of their appearance. They were probably deposited by a glacier or icesheet which had stopped advancing and was beginning to melt, the movements of the ice producing the characteristically elongated shape. Examples can be seen in the Ribble Valley, near Carlisle, and they are numerous in Ireland, continental Europe and North America.

dry adiabatic lapse rate The rate at which temperature is lost when dry air is ascending, normally about 1°C per 100 m.

dry farming A type of farming without irrigation in an area of low rainfall. It is most commonly associated with wheat growing, and examples are found in parts of Australia, southern Africa and the United States. In some dry farming areas the seeds are sown more thinly than usual. This enables each seed to draw water from a wider area. The yield per acre is low, and so dry farming is extensive and farms are normally very large. Crops may be grown in alternate years only, so that the land can be rested. During the fallow year some of the rainfall will be retained in the soil, boosting the following year's total and enabling a satisfactory crop to be grown.

dry site A small elevation in a predominantly wet area. In some marshy environments such as the Fens in eastern England, there are small hills which are suitable for settlements, for example, Ely.

dry valley A river-formed valley which no longer contains a stream. In limestone areas such as the Pennines, there are many dry valleys. They often occur where streams which used to flow on the surface have opened up a joint by solution weathering and have then disappeared below the surface to flow in an underground channel. The chalklands of southern England have hundreds of dry valleys. Some are probably due to the lowering of the water-table; others may be the result of periglacial activity. During the last glacial phase the climate was of the tundra type, and the rocks in southern England were frozen solid by permafrost. The rivers had to flow on the surface, and they managed to erode valleys. When the climate warmed up in post-glacial times and the permafrost had gone, the water sank into the porous rock, leaving streamless valleys on the surface. In some dry valleys in southern England a stream may appear after a period of rain.

dune see **sand-dune**

dust bowl An area which has been changed from a grassy landscape into near desert. If grassland is ploughed and the land left bare, wind can blow away the topsoil and rain will also wash away soil and possibly form gullies. This eroded landscape is bare and useless for farming. It is also possible to expose the soil by overgrazing; if there are so many animals that they remove all the vegetation, wind and rain will erode the soil. There are dust bowls in many places, including east and southern Africa, and the Sahel. The term was originally applied to the dust bowl of the Great Plains in the United States. It is very difficult to repair the damage once the topsoil has been removed, and so in the United States great efforts are made to prevent the creation of further dust bowls. Some of the drier grasslands are not ploughed, cover crops are planted to protect the soil, and there has been some afforestation.

dyke Also **dike 1** An igneous intrusion which is vertical or near-vertical. It is a thin seam of rock which in molten form came from a magma reservoir and forced its way through the existing rock strata. It is discordant to the structure of the other rocks, and is likely to be harder so that after long-term erosion it will form a narrow ridge or slight undulation. Dykes are found in western Scotland, Iceland, and many other present or former volcanic areas. See also **sill** (def. 2). **2** A drainage ditch. **3** An embankment which has been built up to prevent flooding. It may be designed to stop a river from flooding, as in the Fens, or to keep out the sea, as in coastal areas of the Netherlands.

E

earthquake A tremor below the surface of the earth which causes shaking to occur in the crust. The shaking will only last for a few seconds, but widespread devastation can result. The earthquake movements are caused by plate tectonics, and when the shock takes place there are three different waves created; the waves are called P, S, and L (primary, secondary and longitudinal). The P and S waves come from the **seismic focus**, which is the point at which the earthquake originated, and travel up to the surface, where they cause shaking. The P and S waves travel along the surface of the earth as L waves. The point on the surface directly above the seismic focus is called the **epicentre**, and the greatest amount of damage generally occurs near here. A large earthquake is generally followed by a few smaller shocks. If the earthquake occurs beneath the sea, waves can be formed in a way similar to the ripple created on a bond by throwing a stone into the water. These waves are called **tidal waves**, but they are not connected with tides; a better name for them is the Japanese word **tsunami**. Tsunami can travel across oceans at speeds up to 400 km per hour, causing damage thousands of kilometres away from the source of the earthquake. Earthquakes in Chile have been known to cause tsunami which have hit the coastline of New Zealand. One of the largest recorded tsunami was over 10 m high when it hit Java and Sumatra, and the water penetrated hundreds of metres inland. This occurred after the massive eruption of Krakatoa in 1883, and over 30,000 people were drowned. The main earthquake regions are found along the edges of the earth's plates, especially on the perimeters of the Pacific Ocean. Serious earthquakes also occur in the Mediterranean region. They are rare in Britain, and quite small. Earthquakes are usually measured on the Richter scale.

easting The north–south grid lines on ordnance survey maps. They are given first when quoting a map reference.

ebb tide The retreating or falling tide.

eclipse The blocking out of the sun's light by the intervention of a planet between the source and the recipient. When the moon is in a direct line between the earth and the sun, it blocks some of the sunlight which would normally reach the earth; this is an eclipse of the sun. If the earth is in a direct line between the sun and the moon, the sun's rays will not reach the moon; this is an eclipse of the moon.

ecology The science which studies living organisms in relation to their environment and to other living things. It is concerned with the interrelations between the different components of the ecosystem.

economies of scale The reductions in unit cost which result from large-scale as opposed to small-scale production. By means of mass-production and automation it becomes possible to produce goods more cheaply. The aim is always to reduce the unit costs of the items produced. The effects are evident in large integrated motor manufacture.

ecosystem A community of living organisms (plants and animals) and the location or environment in which they live. An ecosystem can be very large; for example, the earth, or very small; for example, an oak or a lawn. All the elements of the ecosystem, whatever its size, will be interrelated. There will be flows of energy and nutrients, and a change in one part of an ecosystem may affect several others. Left to nature, an ecosystem will achieve a balanced state, in which plants and animals live together. A sudden change or a disaster, such as very heavy rain or a prolonged drought, will affect some of the plants and animals, but gradually they will recover. If an oak were blown down, the elements of the oak system would contribute to another system. In recent years many ecosystems have been seriously affected by human activity. There are numerous occasions when the human influence has been harmful, or even disastrous; for example, cutting down trees, ploughing up land, and causing soil erosion. There are major problems which can affect large areas; very large ecosystems are affected by acid rain, and it is possible that the world ecosystem may suffer as a result of the greenhouse effect.

edaphic Relating to soil conditions. Edaphic factors are those which influence the growth of plants, such as soil texture, mineral content, and soil moisture.

edge see under **grit**

E.E.C. (sometimes **E.C.**) see **European Economic Community**

elbow of capture A right-angled bend in a river. It is shaped like an elbow because river capture has taken place. Upstream of the bend is the captured stream and downstream of the bend is the capturing stream. See also **river capture**.

eluviation The removal of material from the A horizon of a soil. Chemicals in solution or small particles of solid matter can be taken down to the B horizon by percolating water. The transported material will accumulate in the illuvial layer.

employment structure The classification of different types of work. Workers are employed in the primary, secondary, tertiary or quaternary sectors. Tertiary and quaternary are often classified together under the heading of tertiary. Primary employment is in agriculture, forestry, mining, and fishing. Secondary employment is in manufacturing industry. Tertiary employment is in retailing, services and administration, and quaternary is in the provision of information, professional advice and expertise. In less developed communities a high percentage of the work force will be in agriculture, but as countries become more advanced and wealthier employment in agriculture declines, manufacturing industry grows, and eventually tertiary and quaternary activities begin to dominate employment. By examining the employment structure figures, it is possible to determine whether a country belongs to the developed or less developed nations.

enterprise zone An area in Britain where special grants and tax advantages are available to anyone creating employment. The government offers incentives to encourage firms and companies to set up in areas of high unemployment. Most enterprise zones are in inner city areas; for example, the Isle of Dogs in London.

entrenched meander see **incised meander**

environment One's surroundings, conditions and circumstances considered as a whole. It may be a natural physical environment or an artificial urban environment. Much of the countryside is also a man-made environment, where agriculture has changed the appearance of the landscape. The soil, vegetation, and buildings are all part of the environment; and pollution, survival of animal species, soil erosion, deforestation, and many other topics are relevant to the continuation of an environment. Conservation groups attempt to influence the ways in which the environment is being changed by man, and forest parks, nature reserves, etc, are being created in many places in order to protect parts of the environment.

environmental lapse rate see **lapse rate**

epicentre see **earthquake**

equator An imaginary line round the earth which represents the 0° line of latitude and encircles the broadest part of the earth. It is 40,076 km (24,901 mi) long and is the only line of latitude which is a great circle, although all lines of longitude are great circles.

equatorial rain forest see **tropical rain forest**

equinox The time of year when day and night are of equal length at all points on earth. The **spring** or **vernal equinox** is on 21st March and the **autumnal equinox** is on 22nd September. On these two dates the sun is directly overhead at noon on the equator, and both the north pole and the south pole will be receiving light.

erosion The process by which the earth's surface is worn away, principally by water (rivers, ice, and the sea) and wind. These agents, together with weathering, are responsible for causing natural changes in the landscape; they are also responsible for deposition, which can create new features.

erratic block A rock which has been transported from its place of origin. Icesheets and glaciers carry large quantities of rocks, sometimes for many miles. Most deposition takes place as the ice is melting, and frequently the rocks are deposited on a completely different type of rock. For example, on the Norber plateau, near Ingleborough, there are many dark Silurian rocks on top of light Carboniferous limestone. There are many other erratics in Britain; for example, Shap granite has been found in southern Lancashire, and Norwegian rocks have been found in north Norfolk. Erratics are also found in many other countries, and in parts of New England and Minnesota they are so numerous that they hinder the use of machinery on the farms.

escarpment A steep slope on one side of a hill of which the other side is a gentle (or dip) slope. The escarpment (or scarp) may be the result of faulting, or, more frequently, caused by tilting of the strata, followed by differential erosion. Good examples of escarpments can be seen in the North Downs, South Downs and Cotswolds. See also **cuesta**.

esker A long narrow ridge formed by a sub-glacial stream. There are often streams flowing beneath glaciers and icesheets, and they carry deposits of sand and gravel. The rivers flow in tunnels and may deposit large quantities of sand and gravel on their beds. The deposits build up and are left behind when the ice melts and the rivers move to other channels. Eskers are usually 10 to 25 m high and 5 to 25 m across; some extend for many miles, especially in Finland and northern Canada. There are many examples in central Ireland, and a few in northern England and Scotland. In Finland there are eskers which form boundaries to lakes, and some are used as routes for roads.

estuary The mouth of a river where tidal effects can be seen. In non-tidal seas, only the effects of salt water will be evident. Most estuaries are funnel-shaped, becoming wider nearer the sea. They may be the result of subsidence and flooding, as on the south Essex coastline. At low tide large expanses of mud or sand are visible in many estuaries, and deposition may be occurring. Some estuaries contain ports, and they provide good sheltered positions, for example London, Liverpool and Glasgow. Dredging may be necessary to keep the channel open, but if the estuary narrows, as at Liverpool, there may be a scouring effect caused by the river flow.

European Economic Community (E.E.C.) An association founded in 1958 by Belgium, France, Italy, Luxembourg, the Netherlands, Western Germany, and since joined by Denmark, Ireland, Greece, Portugal, Spain and the United Kingdom. The community has a common agricultural policy, as well as industrial and trading agreements, and after 1992 trade tariffs between member countries will be abolished in order to allow for the free movement of goods.

evapotranspiration The total loss of moisture by evaporation from water surfaces and the soil, and by transpiration from plants. It is quite difficult to measure evapotranspiration because of the relationships between precipitation, run off, evapotranspiration, and the changes in soil moisture. Precipitation = run off + evapotranspiration ± the changes in soil moisture storage.

evergreen trees Trees which retain their foliage throughout the year, most frequently conifers, except in equatorial regions where, because of the lack of seasonal variation in climatic conditions many deciduous trees become evergreen. As leaves are shed, so new foliage is produced in a continuous canopy. The trees are able to grow new leaves each month because the climate is hot and wet. See also **coniferous forest, deciduous forest, taiga**.

exfoliation A weathering process in which small pieces flake off a rock, rather like skins off an onion. It is most active in deserts and is the result of extreme changes in temperature between day and night. The rocks heat up during the day and expand. As they cool at night they contract. Repeated expansion and contraction weakens the rock, especially if there is water in the rock, as this will increase the amount of expansion. In deserts the presence of dew is a contributory factor. Rocks are rounded off by exfoliation; there may be small piles of rock debris beside the larger rocks. Freeze-thaw activity is similar to exfoliation, but there is a larger amount of expansion involved because water increases by 9% in volume when it freezes. Also called **onion weathering**.

exotic stream The name given to a few large rivers, such as the Nile or the Indus, which can flow across major desert regions because they are fed by rain falling on mountains hundreds of kilometres away. There are more than 50 such streams flowing across the Atacama desert in Peru. See also **intermittent stream, perennial stream**.

exponential (of population) Showing a rapid increase. The graph of world population has shown a dramatic growth during the last 40 years. This increase has been due to improvements in medicine which have reduced the death rate but not been accompanied by a corresponding decline in the birth rate. This rapid increase is described as exponential growth.

exposed coalfield* A coalfield in which coal seams reach or almost reach the surface. In exposed coalfields it is possible to mine by opencast methods, which is cheaper than shaft-mining. However, most exposed

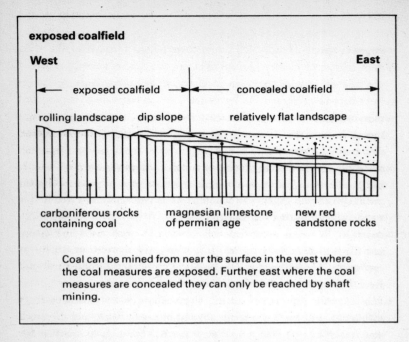

exposed coalfield

West | East

exposed coalfield → | ← concealed coalfield

rolling landscape   dip slope   relatively flat landscape

carboniferous rocks
containing coal

magnesian limestone
of permian age

new red
sandstone rocks

Coal can be mined from near the surface in the west where
the coal measures are exposed. Further east where the coal
measures are concealed they can only be reached by shaft
mining.

fields were utilized in the early days of mining, because the coal was
most accessible. See also **concealed coalfield**.

extensive farming Large-scale agriculture in which large areas of land
may be used, but in which the return per acre is relatively small. Sheep-
farming on hills or in semi-arid areas are a good example of this. Most
extensive farming relates to pastoralism, but there are some areas of
extensive arable; for example, the dry-farming areas in parts of
Australia.

extrusive rocks Igneous rocks which were forced out in a molten state
(magma) at the earth's surface, where they cooled quickly and solidified
as lava. Because they cooled rapidly, extrusive rocks contain small
crystals, which may be too small to be seen by the naked eye. If magma
cools and solidifies beneath the surface, cooling will be slower and
larger crystals will form. Rocks formed in this way are called **intrusive
rocks**. See also **magma**.

F

factory farming Intensive livestock-farming, in which cattle or pigs are kept permanently indoors and fed a carefully regulated amount of food each day. The animals are not allowed to walk about very much, as too much exercise would use up energy and slow down their growth rate. It is a very commercial type of farming, which aims to produce a good quality product as quickly as possible. When chickens are reared in this way it is called **battery farming**.

Fahrenheit A temperature scale which was commonly used in England and English-speaking countries. It is now being replaced in Britain by centigrade. Freezing and boiling points of water on the Fahrenheit scale are 32° and 212° respectively.

fallow A term denoting farmland which has been left unseeded for a season or more in order to rest it. The land may be tilled, and the weeds may be killed off during the fallow period. Some fields may be left fallow for a season before planting a late or winter crop. In some tropical areas bush fallow is practised; with this system the land may be left fallow for several years.

family planning Birth control by the use of contraceptives. This is already common practice in many advanced countries, and is now an urgent requirement for most of the Third World countries, where many of the social and economic problems result from over-population in relation to the available resources.

famine An extreme and general shortage of food. Famines have been an occasional problem for hundreds of years, generally occurring as a result of variations in the amount of rainfall, or natural disasters such as volcanoes and hurricanes. Too much rain can cause flooding and delay planting of crops, or destroy crops which were already growing. Too little rain will prevent crops from growing, and so the harvest will be small. Although there is now more food grown in the world and more help available for people who are starving, populations have grown so rapidly that famines occur more frequently than in the past. In some countries, such as Ethiopia, famine is a common occurrence.

farming system Any type of farming with its own distinctive characteristics, and inputs and outputs. A dairy farm in Cornwall is an example of a farming system, and a cereal farm in East Anglia is a different farming system, with more machines, bigger fields, different products. A shifting cultivator in the Amazon is a totally different system, as is a Lapp reindeer herder; there are many other examples all over the world.

farmscape A landscape which has been created by the farmers. On the borders of England and Wales there are many small fields with numerous hawthorn hedges, which creates a very different landscape from the Canadian prairies. Farming methods determine the appearance of the landscape. A recent change to farmscapes in Britain has been the mass removal of hedges to create large fields.

fault* A crack or fracture in the rocks of the earth's crust with an associated movement of the strata on either side. The movement will be slow and quite small, only a few centimetres, though fault movements often continue for thousands of years. In such cases uplift or downthrow of hundreds of metres is possible. Faulting is caused by plate tectonics, when movements in the crust create stress and tension in the rocks, causing them to stretch and crack. Vertical movements cause normal and reverse faults, and horizontal movements cause tear faults. The vertical change of height on opposite sides of a fault is called the **throw**, and the horizontal movement is called the **heave**. Faults often occur in groups and if two or more roughly parallel faults cause a block of land to rise, it would be called a horst or block mountain. If the land sinks in between two or more parallel faults, it would create a rift valley or graben.

fault plane The surface along which faulting has occurred. In the case of a normal fault, the fault plane will be vertical or it will be inclined so that the downthrown side is on the dip side of the fault plane.

fault scarp A scarp which is located along the line of a fault. Giggleswick scarp in the Pennines is an outstanding example.

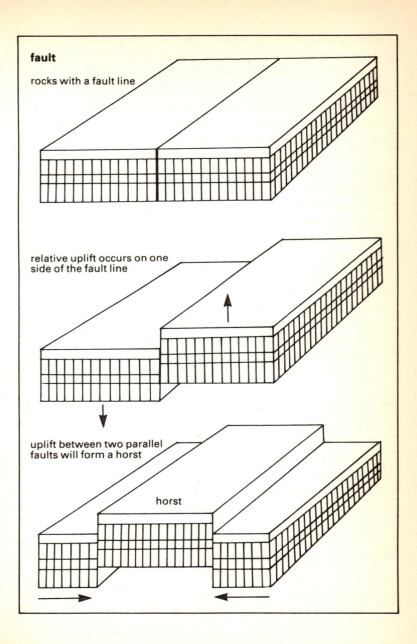

fault

rocks with a fault line

relative uplift occurs on one
side of the fault line

uplift between two parallel
faults will form a horst

horst

favela (in Portugal and Brazil) A spontaneous or shanty settlement which has grown up in a large city such as São Paulo.

feed Food for animals. It may be hay, root crops, alfalfa, oilseed cake, etc.

fell An upland in northern England, especially in the Pennines and the Lake District. The fells are areas of rough grazing often used for sheep in the summer. Some of the fells are common grazing, and some are important grouse moors. See also **hill farming**.

felspar Also **feldspar** Any of a large group of minerals which are silicates of aluminium, with some potassium, sodium, calcium, or barium. Almost half the earth's crust is made of felspars. Common varieties include **orthoclase**, which is often whitish in colour, and **plagioclase**, which is often pinkish; both varieties can be found in granites.

fen A flat low-lying area which is marshy. Regions of fen occur in many river valleys where there used to be lakes, and near the coast where land has been reclaimed from the sea by silting. The Fens are an area near the Wash in eastern England, where the marshes have been drained and the former wetlands have been turned into very arable farmland. There is another large expanse of fenland in Somerset, the Somerset Levels. Some fen soils are very peaty and acidic, but others are alkaline in character.

Ferrel's law The law that all moving bodies on the earth's surface will be deflected to their right in the northern hemisphere and to their left in the southern hemisphere, because of the rotation of the earth. This is especially apparent with reference to winds, but it can also affect ocean currents, and people; for example, people who are lost in fog or in a desert walk round in circles as a result of Ferrel's law.

fertilizer Any material used to improve the fertility of the soil. Natural fertilizers are obtained from animal manure, rotting vegetation, fish bones, etc. Many artificial fertilizers are now produced by chemical industries, using potash, phosphates, and nitrogen.

fetch The extent of sea across which winds can blow. If there is a large expanse of open sea, it is likely that high waves can be expected. For example, waves which come in from the Atlantic are likely to be larger than waves which form in the narrow parts of the English Channel.

finger lake A long narrow lake situated in a glacial U-shaped valley. As the glacier moved downhill, deepening and widening the valley, the irregular movements of the ice along the valley floor would cause uneven erosion. When the ice melted, hollows would be left behind in rock basins. The steps or riser at the end of each basin would form a dam, and occasionally there would be deposits of glacial drift to add to the damming effect. Ribbon lakes are numerous in the Lake District (for example, Windermere, Ullswater), in Scotland (for example, Loch Ericht), and in north Wales (for example, Llyn Peris). All these lakes will gradually silt up because of sediment brought in by the rivers flowing through the lakes. Also called **ribbon lake**.

fiord see **fjord**

firn (*German*) Granular snow, halfway between snow and ice, which is subsequently compressed and compacted to form ice. Also called **névé**.

fissure A cleft or crack in a rock, especially one which is the site of a fissure eruption.

fissure eruption A volcanic eruption in which lava pours out of a fissure and continues to flow for a few hours, sometimes days. Basic lava is fluid and free-flowing, and can travel across large areas of countryside. There may be several eruptions along the same fissure line. When this happens, large deposits of lava can accumulate, and there are examples of plateaus which have been formed in this way, such as Snake-Columbia plateau in the northwest United States, parts of the Deccan Plateau in India and the Antrim Mountains in northeast Ireland.

fjord Also fiord An inlet of the sea which was formed by glacial erosion. It is really a flooded U-shaped valley. Fjords are generally narrow and steep-sided, and the water can be hundreds of metres deep. The bed of the fjord is often undulating because of the irregular curve of glacial erosion. Sometimes there is quite shallow water at the seaward end, because less erosion took place near the end of the glacier. Large examples of fjords can be seen in Norway, southern Chile, and South Island, New Zealand, and there are some examples in northwest Scotland and northwest Ireland.

flash flood A rapid flood, often quite dramatic, caused by heavy rain followed by a rapid run-off. Flash flooding is most likely to occur in deserts, where there is no vegetation to slow down the rate of run-off, but it can occur in wetter areas too. Heavy rain, probably the result of a thunderstorm, may fall several kilometres away from the site of the flood. Routes followed by flash floods are sometimes well known, and in the United States and Israel there are places where signposts give warnings of the likelihood of flash floods.

flint A hard rock which consists mainly of the mineral silica. Flint was formed from the remains of dead sea-creatures whose bodies were of silica. It is often found in chalk, which is also derived from dead sea-creatures, but of a different kind. Layers of flint commonly occur in the chalklands of southern England. They are a nuisance in fields, because they are so hard and interfere with machinery. They are often used for building purposes, as they are very resistant to weathering. Many churches in chalkland regions have been faced with pieces of flint.

flooding Flowing of water over land which is not usually submerged. Very heavy rainfall, prolonged steady rain, or melting snow can be the cause of flooding. Some river basins, for example, in granite areas, have a rapid run-off, and water quickly reaches the river; in limestone and chalk areas, the water passes slowly through the rocks and the danger of flooding is reduced. In porous rocks the peak flow in rivers often comes a day or more after even the heaviest rainfall, whereas in impermeable areas the peak flow is only hours after rainfall. A good cover of vegetation, especially with woodland, delays the throughflow of water. If the vegetation cover is removed, there may be a much greater danger of flooding. Many parts of the world, including Nepal, Thailand and the Amazon basin, now suffer from more serious flooding than in the past because water runs off far more quickly as a result of extensive tree felling in these areas.

flood plain* A lowland alongside a river which has been built up by the deposition of alluvium in times of flood. Many flood plains are only a few metres in extent, but on some rivers, such as the Mississippi and the Amazon, the flood plain can stretch for several kilometres. The flood plain of the river Hwang in northern China is over 160 km wide, and there are large levées alongside the river. When the river floods, the

flood plain

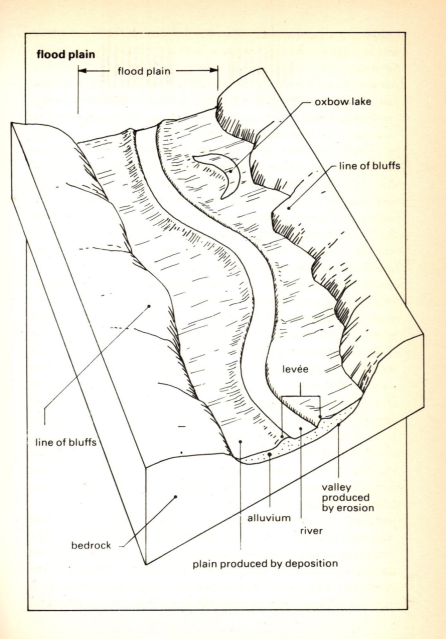

flood plain

oxbow lake

line of bluffs

line of bluffs

levée

bedrock

alluvium

valley produced by erosion

river

plain produced by deposition

water cannot always return to the main channel, and so a new channel may be created; in this way the Hwang has frequently changed its course.

flow line A type of diagram on which a line represents the movement of goods or people. The width of the line is proportional to the amount of movement which is taking place. To show the international trade in wheat, for example, arrows could be marked on a world map; there would be thick arrows from the United States and Canada, and also an arrow from Australia to show the main exporters. The arrows would then split up to indicate the main wheat importers.

fluvial Relating to a river or a stream.

fluvio-glacial Relating to a river which has been formed by meltwater from a glacier or icesheet. There are often rivers which flow beneath glaciers; they may form eskers and kames, which are fluvio-glacial deposits. At the end of an icesheet, where melting is taking place, there is likely to be a large quantity of water flowing across the plain or down the valley. At first it may be a sheet flow, but gradually it will become a channelled flow. It will pick up and transport particles of sand and perhaps gravel which have been dumped by the ice. Coarse particles will be dropped first, but the fine clays may be carried considerable distances. The deposition of the sediments will be according to weight and size, and so sorting inevitably takes place; fluvio-glacial deposits are always sorted, and often stratified, too. The bulk of fluvio-glacial deposits are found in the outwash plain, which is quite close to the downstream end of the ice.

fog A mass of small water droplets floating in the air which make visibility less than one kilometre. If visibility is more than one kilometre, it is mist. Fog is caused by water vapour condensing as a result of the air becoming cooler. It is most likely to form in settled anticyclonic conditions. After a hot and sunny day, the air is warm. When night falls the air cools, and so is less able to hold water vapour. The water vapour condenses to form millions of tiny droplets. The coldest air is likely to roll downhill into valleys, which is where fog frequently forms. Fog is also quite likely to form first near rivers or lakes, where there is a greater supply of moisture. If a wind develops, the turbulence will clear the fog. Fog will also clear in the heat of the sun during summer days. In winter, however, it can persist throughout the day. **Hill fog** is really low cloud, and is associated with low pressure. It may persist for several hours, but will not affect low-lying areas.

föhn A warm dry wind which blows down the valleys on the north side of the Alps. It is associated with the movement of a depression to the north of the Alps. The low pressure of the depression draws in air from the south, which rises over the Alps; as it ascends condensation occurs and there is rainfall. The ascending air loses heat at a slow rate because the air is wet. Once over the range of mountains, the air is dry and so gains heat at a more rapid rate. By the time it descends the valleys on the other side it has become a very warm dry wind. The föhn occurs most frequently in spring, when it is useful for melting the snow. If it occurs in autumn it can help with ripening the grapes, as an increase in temperature of 10°C is often recorded. A similar wind which blows over the Rockies in Canada and the United States is called the chinook.

folding* The process by which rock strata, under pressure as a result of movement of the earth's plates, buckle and bend. If the compression is fairly gentle and quite even from both sides, the rocks may fold into gentle and symmetrical shapes. If the pressure is uneven, then asymmetrical folds will form. In many cases, the folds are pushed right over to form recumbent folds and overfolds. If a large fragment of land is broken off by the effects of folding, it is called a **nappe**. Nappes are very numerous in the Alps, where intense folding has taken place. Simple folds are found in the Jura mountains, and a good example of an asymmetric fold is found in the hills of the Isle of Wight. See also **anticline, syncline**.

fold mountains Mountains formed as a result of folding, including the Alps, Andes, Himalayas and Rockies. Convection currents beneath the earth's crust cause the plates to move and compress the sediments; the sedimentary rocks are then folded and uplifted to form mountains. In association with the folding there is sometimes volcanic activity, and the central core of some fold mountains contain igneous rocks. There have been three major periods of folding and mountain formation during the last 500 million years. The old fold mountains were formed during the

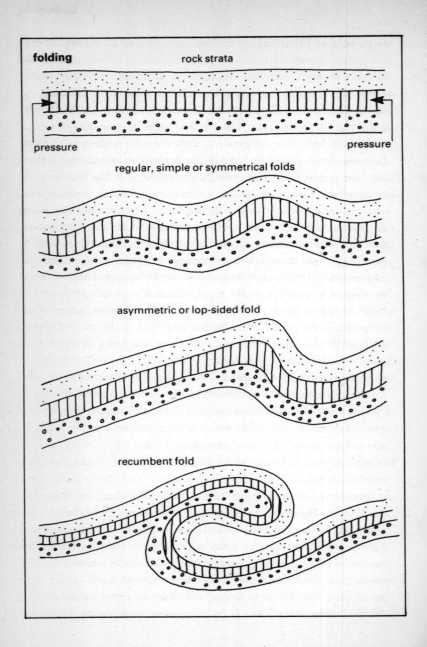

folding

rock strata

pressure · · · · · pressure

regular, simple or symmetrical folds

asymmetric or lop-sided fold

recumbent fold

Caledonian and Hercynian period and the young fold mountains during the Alpine period. Caledonian folding occurred at the end of the Silurian, and formed the mountains of Scotland and Scandinavia. Hercynian or Armorican folding came at the end of the Carboniferous period and formed the Pennines, some of the Welsh mountains, the upland of Brittany and the Harz mountains in Germany. Alpine folding took place during the Tertiary era, and, since the processes of erosion have been at work on these mountains for only 40 years, they are not only the youngest but also the highest mountains in the world

footloose industries Light industries which have few basic requirements and are therefore able to set up in almost any location offering an adequate power supply and access to an efficient means of transport.

fossil fuel Any of the non-renewable fuels formed from the remains of dead plants and animals which accumulated in the ground: coal, oil, natural gas and peat. They formed very slowly but are being consumed very quickly, and are likely to be exhausted within the next few hundred years. Burning fossil fuels has led to much pollution of the atmosphere and to the phenomenon known as acid rain. Increasing attempts are now being made to clean up power stations, so that they do not give off as many harmful gases.

free port A port or a zone within a port where no customs duty is paid. Designed to attract more customers and more trade, many free ports, such as Hong Kong, are entrepôts, which handle goods to be passed on to other countries. Some free ports are located at airports; for example, Manaus in Brazil. Also called **free-trade zone**.

freeze-thaw A process in which extreme changes of temperature contribute to the break-up of rocks. During a cold spell, probably at night, the water particles in rocks freeze. When this happens the water increases by about 9% in volume, and this places a great strain on the rock. As the temperature rises, the water melts. If this process is repeated day after day for many years, the constant expansion and contraction will gradually break up the rock. Freeze-thaw occurs in mountainous areas and tundra. Loose rock fragments resulting from freeze-thaw activity can be seen in the Alps and in the higher parts of Britain. In glacial areas it is a major contributor to the erosion of backwalls of corries and other exposed rocky outcrops.

frequency curve A graph which shows the frequency, or number of occurrences, of any chosen item, such as temperatures, answers to a questionnaire, etc. The values are often grouped into classes so that a histogram can be drawn. On a frequency curve graph, the horizontal axis gives a scale for the range shown by the variable, and the vertical axis gives the frequency of each variable, either as an actual figure or as a percentage.

frequency distribution The range of values covered by any set of data, generally shown as a bar graph or histogram. A population pyramid is a frequency distribution diagram.

friction of distance The restricting effect of distance on human activity and general accessibility. The amount of traffic tends to decrease with increasing distance from a big city such as London. The number of walkers on a footpath decreases in proportion to the distance from the nearest car park. In trading goods, an increase of distance is likely to increase the cost, and therefore decrease the likelihood of selling the commodities. The general formula for friction of distance is:-

$$f = a\ \frac{1}{d^2}$$

f is the volume of movement
d is the distance
a is the constant

There are several useful gravity models which consider the effects of distance. However such models cannot really consider the effects of developments in transport, which quicken up the flow in some cases, and cannot take into consideration the peculiarities and idiosyncracies of people.

front A dividing line between two air masses. See also **cold front, occluded front, warm front**.

frontal rainfall Rainfall which is formed in a depression, where the fronts cause a far larger and more rapid uplift of the air. Large quantities of air rise quite quickly at the fronts and the rainfall can be heavy, though it is unlikely to last for more than a few hours.

frost Frozen droplets of water. When the air temperature falls, generally at night, water vapour may condense. If the air temperature then falls below zero, the water will freeze, forming a covering of **ground frost** on the ground and on any solid exposed objects. If the temperature falls a few degrees below zero, it is possible for the water vapour in the air to freeze to form **air frost**, but this is not as common as ground frost; both types of frost are called **hoar frost**. Frost is most likely to occur in anticyclonic conditions when the air is fairly calm, and there is little wind or turbulence to mix up the layers of air. The air is cooled by radiation from the ground.

frost hollow A valley bottom or basin which is prone to frost. In a frost hollow cold air accumulates, producing lower temperatures than those on adjacent slopes; night minimum temperatures may be as much as 4° to 5° lower than in neighbouring areas. Any turbulence will cause the air to mix with the air above, and so frost-hollow effects are noticed only in calm anticyclonic conditions. One of the greatest frost hollows in the world is at Verkhoyansk in Siberia, where January average temperatures are about −45°C. In frost hollows the coldest temperature is just above ground level; above that the air temperature is higher. Normally air temperatures decrease with height, and this converse situation in frost hollows gives rise to temperature inversion, with a warmer layer above a zone of colder air. The inversion layer is likely to form cloud, which will restrict the upward movement of dust or smoke. Pollution is likely to accumulate below the inversion layer, and this is one of the contributory factors to the fogs of Los Angeles and the former smogs of London.

Fulani A large group of nomadic and semi-nomadic pastoralists living in northern Nigeria and parts of adjacent countries. They normally have a permanent village, to which they return for a few months each year, but they have to wander with their herds of cattle for several months each year, in order to find pasture.

functions Collectively, the services, amenities, and goods available in a settlement. A large city has many more functions than a small village. Low-order functions are those which provide convenience goods likely to be required frequently, perhaps every day, such as newspapers, bread and milk, and these will be found in most settlements. High-order functions, which supply specialized or long-lasting goods, will be available only in larger towns.

G

gabbro A coarse-grained igneous rock. The mineral content of gabbro consists mainly of felspars and ferro-magnesian minerals; for this reason it is darkish in colour and alkali in content. The coarse grains indicate that it was formed at depth beneath the surface, and was able to cool slowly, allowing time for the formation of large cyrstals. The equivalent acidic igneous rock with large crystals is granite.

gap An opening in a ridge, often created by a river. Some gaps are formed by glacial erosion, or by the effects of melt-water at the end of a glacial phase. Gaps often become important route ways; for example, Cumberland gap in the United States, which was followed by many early explorers, and Tyne gap in northern England. The ease of passage through gaps often resulted in the development of settlements nearby. Many villages and towns can be described as gap towns; for example, Dorking on the river Mole, and Arundel on the river Arun.

gentrification The improvement of houses and the creation of new and often expensive houses in some inner city areas which have suffered from industrial decline. In such areas there is often a shortage of cheaper housing to suit the needs of local people, and this has created some problems in places such as Tower Hamlets and other parts of London's old docklands, where the new housing has been bought up almost entirely by outsiders.

geographical inertia The tendency of a place to continue with a particular industry or business activity after the original advantages afforded to that activity have disappeared. This is because the factory, machinery, local labour supply, tradition and reputation may allow the industry to continue successfully. Such is the case with the pottery industry in Stoke on Trent. Originally, the industry was established there because there were local supplies of clay and coal; nowadays the clay is brought in from other areas, but the industry still continues. A similar situation applies to the engineering works of the Black Country. Also called **geographical momentum, industrial inertia.**

geography The study of the earth's surface and its people. The two main sub-divisions are physical and human geography. Physical geography includes geomorphology, climatology, meteorology, and pedology; and human geography includes the study of agriculture, industry, resources, and political geography. Geography also involves the study of such varied present-day topics as conservation, racialism and traffic congestion, and is closely linked with many relevant issues of modern life.

geological column A diagram which shows the geological time scale. Also called **stratigraphical column.**

	length of period (in millions of years)	time the period began (millions of years ago)		
Pleistocene	2	2	Quaternary	
Pliocene	10	12		
Miocene	13	25		Cainozoic
Oligocene	15	40	Tertiary	(era of recent
Eocene	20	60		life)
Palaeocene	10	70		
Cretaceous	65	135	Secondary or Mesozoic	
Jurassic	45	180	(era of middle life)	
Triassic	45	225		
Permian	45	270		
Carboniferous	80	350		
Devonian	50	400	Primary or Palaeozoic	
Silurian	40	440	(era of early life)	
Ordovician	60	500		
Cambrian	100	600		
Pre-Cambrian	4000			

The dates for the geological periods are based on radiometric dating, but should not be regarded as absolutely precise. The older periods, especially, may vary by a few million years.

geological time scale The sequence and approximate duration of the geological periods.

geology The study of the origin and structure of the earth and of the changes it has undergone and is in process of undergoing. Geologists work from the crust inwards, whereas geographers look at what there is on the surface.

geometric rate A sequence in which the ratio of any number to its predecessor is a constant; for example, 2, 4, 8, 16, 32, etc. Thomas Malthus believed that world population would increase at a geometric rate, while food supplies would increase at an arithmetic rate.

geomorphology The study of the shape of the earth. Geomorphology considers the relationships between geological structures and surface landscape features, as well as the processes which change the features by erosion and deposition. Rivers, coastlines, rock types, slope formation, ice, erosion, and weathering, are all within the scope of the geomorphologist. Also called **physiography**.

geosyncline A large syncline which may extend for hundreds of kilometres. It is a basin-shaped area located between two of the earth's plates. As the plates move closer together, the geosyncline is compressed. Sediments which have accumulated on the floor of the geosyncline will be crumpled and buckled to form fold mountains. The Tethys Sea, an enlarged version of the Mediterranean, was a good example of a geosyncline; as it was reduced in size, the Alps and Atlas mountains were formed.

geothermal Relating to the heat of the earth's interior. Geothermal activity produces geysers, springs, mud volcanoes, and steam. Some of these phenomena have now been harnessed to provide energy. New Zealand, Iceland, Italy, and California are important producers of power of this kind. In Iceland cheap heating is provided to the capital, Reykjavik, using hot water from underground. There are also many greenhouses which are warmed by underground heat. Experiments have begun to try to produce heat from the granite rocks of southwest

England. Geothermal heat is a clean form of energy, and it is renewable.

geyser A natural spring which intermittently throws out jets of water and steam. Geothermal heat within the rocks, generally in areas which have had volcanic activity, warms up the ground water. Water which accumulates in a tube or vent can warm up to such an extent that the water near the bottom of the vent begins to boil, causing the water higher up to blow into the air as a natural fountain. The erupted water falls back to earth, trickles back into the vent, and the warming process begins again. The name geyser is derived from the Icelandic *Geysir*, and the Great Geysir used to erupt every 30 minutes or so. It is now nearly extinct, but can be made to erupt, for the benefit of tourists, by feeding it with special additives. There are numerous other geysers in Iceland which do erupt naturally. The best known geyser in the United States is Old Faithful in Yellowstone National Park, which used to erupt every 66 minutes, but it too has become less reliable and quite irregular.

glacial Relating to a glacier. **Glacial advance** is a period of increasing snow and ice, when glaciers move down valleys. **Glacial retreat** is the reverse process. Less snow and ice means that the melt at the snout of the glacier is more than the replacement rate from the snow field at the top of the glacier. As the glacier shrinks, it appears to be retreating. Glacial conditions are those experienced during a glacial phase, when sub-zero temperatures continue for most of the year. Much of Greenland and most of Antarctica are experiencing glacial conditions at present. Britain has experienced glacial conditions several times during the last two million years, but there have also been warmer interglacial periods, such as the present period, which has been in existence about 12,000 years.

glacial phase see **ice age**

glacier A mass of ice which flows down a valley. Snow accumulates on mountainsides, above the snow line, but gradually starts to move downhill because of the effects of gravity. As the snow is compressed by the weight of more snow on top of it, it is likely to melt and then refreeze into granular ice called firn. Continued pressure gradually turns the firn into ice. As it moves downhill, a glacier normally follows the route of an existing river valley, which is likely to be deepened, flattened, widened and steepened by the action of the ice. The glacier will grow and extend further down the valley, as long as the supplies of snow and ice from the snow field are greater than the rate of melt at the lower end of the glacier. If a glacier continues to grow it is said to be advancing, but if it shrinks it is said to be retreating. There is a seasonal advance in winter, and a retreat in summer, whatever the long term changes. During the glacial periods, many glaciers extended down the valleys and on to the lowlands beyond. The ice spread out on the plains to form ice sheets.

glasshouse see **greenhouse**

glei Also **gley** A type of soil which occurs in wet areas. Where the water-table is near the surface and drainage is very poor, waterlogging will be the result. The soils will be greyish in colour and probably contain some brownish deposits of ferric hydroxide. Glei soils are found in tundra areas and in wet, badly drained hollows in more temperate zones.

gneiss A metamorphic rock with a distinctive layering or banding. Because of heating during metamorphism, the different minerals melted and then solidified in distinct layers. The darker minerals are likely to be hornblende, augite, mica, or dark felspar. Before metamorphism, gneiss was an igneous rock, possibly a granite. Good examples of gneiss can be found in northwest Scotland, especially on the island of Lewis.

Gondwanaland see **continental drift**

gorge A narrow steep-sided valley. Gorges are found in hard rock; otherwise the steep sides would be rapidly eroded. Good examples can be seen in areas of carboniferous limestone, such as Cheddar and Gordale. Canyons are examples of large gorges, and the largest examples of these can be seen in the southwest United States.

graben (*German*) see **rift valley**

graded profile A smooth profile of the course of a river. It occurs when all irregularities such as waterfalls have been eliminated. The longitudinal profile of a river valley is a smooth curve, whereas the long profile of a glacial valley will have irregularities.

gradient 1 A slope which can be shown on ordnance survey maps by the contours; the closer the contours the steeper the gradient. In order to calculate the average gradient between two points, it is necessary to work out the difference in height between the points, and to measure the distance. Difference in altitude divided by the distance apart will give the gradient; for example,

$$\frac{500}{5000} = \frac{1}{10}$$

This gives a gradient of one in ten (1:10). 2 A curve on a graph used to represent a rate of change in relation to distance, as for example the decline of population density from the centre of a city outwards.

granite A coarse-grained acid igneous rock. It consists of large grains visible to the naked eye, which means that it was formed beneath the surface and cooled slowly. This enabled the liquid magma to solidify into large crystals. Granite is light in colour because of its high acid content, and its main minerals are quartz, some felspars, and micas. Granite is a hard rock and slow to weather. Although formed underground, it is often seen on the surface because all the other overlying rocks have been eroded. For this reason, too, it generally forms upland regions, such as Dartmoor, Bodmin Moor. Weathered granite disintegrates, leaving kaolin as a residual mineral.

grassland A region in which the natural vegetation is mainly grass. Grasslands occur in areas where the rainfall is too heavy for a desert but insufficient to support a forest. The total annual rainfall is likely to be more than 250 mm, and as much as 1000 mm. The seasonal distribution may be important, because in some tropical areas where evaporation is very high the effectiveness of summer rainfall is very low. For this reason some locations with 1000 mm rainfall are unable to support woodland. There are numerous varieties of grass, but two main types of grassland; these are the temperate and the tropical grasslands. Temperate grasslands are found in central North America, where they are called prairies, and in the Soviet Union, where they are called steppes. An extension of steppe conditions can be found in Hungary. There are also temperate grasslands in the southern hemisphere: on the pampas of Argentina, the veld of South Africa, the Murray-Darling basin in Australia, and the Canterbury Plains of South Island, New Zealand.

The steppes and the prairies have a continental type of climate, with warm summers and summer rainfall, but dry and very cold winters. Temperature ranges from 15° to 20°C in July, and −10° to −20°C in January, with a total precipitation of 500 mm. The southern hemisphere grasslands are monsoonal or eastern marginal and they are located in narrower land masses. Winters are milder, 5° to 10°C; summers are about 20°C, and the rainfall is heavier − 500 to 750 mm − and is less concentrated in the summer months. The grass in the temperate areas is fairly short, remains greenish throughout the year, and provides good pasture for animals, although much has now been ploughed up for cereals. Tropical grasslands, or savannas, have very tall grass, up to 2 m, in the hot wet summers, but in the hot dry winters it goes brown and shrivels. The total rainfall may be 1000 mm, but most falls in summer, when evaporation is high. Summer temperatures are 25° or more, and winters are about 15° to 20°C. Tropical grasslands are found in east Africa, Nigeria, northern Australia, the Orinoco basin, and parts of the Brazilian plateau.

grass minimum thermometer A thermometer used to record the minimum temperature 2 to 3 cm above ground level, where the air temperature often reaches its lowest point. A metal rod reveals the lowest temperature experienced during the night.

gravel A coarse sedimentary rock, often unconsolidated. Gravels may be cemented, to form conglomerate if the particles are rounded, or breccia if the fragments are angular. Unconsolidated deposits may be found on river terraces or in areas of glacial drift.

gravity model A model which analyses or studies some kind of movement. It is generally used for investigating and attempting to predict the movements between two particular locations. It is thought likely that the population movements between two settlements will be proportional to their size; also that the attractions of two settlements to the population of a third, nearby settlement will be proportional both to the size of the two and also to the distances at which they are situated. The simplest type of gravity model utilizes the following formula.

$$Mij = \frac{PiPj}{(dij)^2}$$

M is the volume of movement between the two settlements i and j. Pi and Pj are the populations of the two settlements, and dij is the distance from i to j.

See also **Reilly's Law of Retail Gravitation**.

great circle An imaginary circle drawn round the earth. If a cut was made through the centre of the earth along the line of a great circle, it would split the earth in half. The equator and all lines of longitude are great circles, and so too are numerous diagonal lines. The shortest distance between any two points on the surface of the earth, will follow the line of a great circle. Sailing vessels used to follow great circle routes, and nowadays many air routes, especially over the Arctic, follow the line of a great circle.

green belt An area of green mainly rural land surrounding an urban area. The major green belt is the metropolitan green belt, which surrounds London, but there are other green belts round other major cities. Over 10% of England is designated as green belt. There are strict controls over building and development in green belt areas, although agricultural buildings may be constructed. Planning permission is necessary before any kind of urban development takes place, and urban sprawl is prevented from spreading into green belt areas.

greenhouse A building, mainly of plastic or glass, for the cultivation of plants. Greenhouses are used to create an artificial climatic environment in order to protect or grow crops in the winter, or produce earlier vegetables than is possible out of doors. Some greenhouses are heated to enable growth throughout the year.

greenhouse effect The warming of the lower layers of the atmosphere as a result of retention and reradiation of solar heat. As the sun's rays pass through the atmosphere on the way down to earth, some heat is absorbed but most of the short-wave solar energy passes through. This energy is given off by the earth as long-wave radiation, which cannot easily pass up through the atmosphere. The heat is retained more easily in the atmosphere if there is a cloud layer. More heat is retained now than in the past as a result of burning fossil fuels, which produces millions of tiny particles; the particles rise into the air, forming an extra layer which acts rather like the glass in a greenhouse, and traps the heat, causing the climatic conditions to warm up slightly. In the next few decades the process may continue, causing a considerable melt of Antarctic and Greenland ice, which would cause a rise in sea-level and the flooding of several of the world's major cities, including London and New York.

Green Peace A conservation group, which is active in many parts of the world. It will support any kind of protest its members consider relevant to their ideas of protecting and preserving wildlife. As a group they are strongly opposed to the construction and use of nuclear power stations and nuclear weapons.

green revolution A period when rapid progress was made in the production of extra food supplies. In the 1960s research at the Rockefeller Centre in Mexico produced new strains of wheat and maize which could produce higher yields per hectare. At about the same time, research in the International Rice Research Institute in the Philippines produced several new rice hybrids. Suddenly the world was able to produce more food. The development of new hybrid varieties is a continuing process; at the same time, new pests evolve, and so it is a constant race to keep ahead of all the problems.

grid A pattern of vertical and horizontal lines drawn on a map in order to provide a basis for map references. On ordnance survey maps the British National Grid is used, and numbering of the grids (vertical and horizontal) starts in the bottom left-hand corner of the map. The vertical grid lines are used to determine the easting, that is, the location of a point to the east of the nearest grid line; while the horizontal lines determine the northing, that is, the location of the point to the north of the nearest grid line.

grid iron drainage A drainage system in which there are a number of parallel streams, each with tributaries that run at right-angles to the main stream, so that there exists a rectangular network of streams. Examples can be seen in the Appalachians, the Jura mountains and the Weald in southeast England. Also called **trellis work drainage**.

grid iron pattern A pattern of roads shaped like a gridiron. The roads run parallel to each other, with a second set that cross at right angles. This rectangular plan which is common in many towns in the United

States, is often the result of careful planning before much building had taken place. Most European towns were not planned, and therefore have a more irregular plan, although the central parts of some towns and some sections of new towns show a gridiron pattern.

grike see **gryke**

grit A coarse type of sandstone in which the grain size is larger than in a typical sandstone. A good example of grit is the millstone grit which is found in many parts of the Pennines. It gives rise to bleak moorland, with large expanses of peat bog, and there are often small but steep rocky outcrops called **edges**.

gross domestic product (G.D.P.) The total value of goods and services produced by a country over any particular period, normally a year. All sections of the economy contribute to the G.D.P. If the total gross domestic product is divided by the total number of the population, a G.D.P. per capita figure can be discovered. This is a useful general guide to the wealth and standard of living of a country.

gross national product The gross domestic product plus the money earned by investments abroad, but minus the money earned by foreigners living in the country.

ground frost see **frost**

ground water Water which is found underground, in the pores and cracks of the rock. It consists mainly of rainwater, which has percolated down into the rocks. Most ground water is near the surface, and, if it reaches the surface, it will cause a puddle or a lake. Ground water moves downwards because of gravity; it may move sideways through rocks, especially where an impervious stratum prevents downward movement. In some rocks there may be considerable movement of water, and this is particularly marked in carboniferous limestone. Underground streams can sometimes be seen, which have created many miles of tunnels and caverns. As it moves, the water can cause erosion and chemical solution; many chemicals may be dissolved from the rocks by the effects of ground water.

growing season The period in which plants normally grow. In Britain the growing season extends from May to September or October, between the last severe frosts of spring and the first severe frosts of autumn. The growing season becomes shorter with altitude, and for this reason cereals are not grown on Dartmoor or the Pennines. Only grass can survive on hills and so there is usually pastoral farming, but not arable. In central United States the corn belt is commensurate with the area which has 150 frost-free days. If the growing season is shorter, wheat is the major crop; if the growing season exceeds 200 frost-free days, cotton is grown.

growth pole A town or group of towns in which industrial development and expansion is taking place. In advanced countries it is possible to set up industries in a wide range of locations, because transport, water, electricity (infrastructure) are widely available. In poorer countries there are few locations where developments are possible, because of the lack of infrastructure, and so most industries congregate in the capital, main port, or one of the few large towns. In order to help developments in less favoured areas, it is important to create new centres of growth, or growth poles. In Brazil, the development of the new capital, Brasilia, created a new growth pole. In Venezuela, the developments in the Orinoco and Caroni valleys near Ciudad Guayana are another example of a growth pole.

groyne A low barrier of concrete or wood built out into the sea from a seawall or promenade. It slows down the movement of longshore drift, and also protects the seawall from the direct battering of the waves. The length and frequency of groynes depends on the angle at which the prevailing waves come in to the shore. Many coastal towns erect groynes in order to protect the coast and save money on sea defences, since groynes are more cheaply repaired and replaced than buildings or roads.

gryke Also **grike** An enlarged joint in carboniferous limestone. Where solution has weathered away at a joint, a vertical crack is opened. Grykes may be a metre or more in depth and there are generally several of them in close proximity. Down in the sheltered hollows created by the grykes, there is often a wealth of small lime-loving plants. Good examples of grykes can be seen near Malham and in many other places in the Pennines. See also **limestone pavement**.

Gulf Stream An ocean current which flows out of the Gulf of Mexico and northwards along the east coast of the United States. As it move northwards it begins to turn towards the right, because of the effects of the rotation of the earth. It then begins to cross the Atlantic; at this point it is more correctly known as the North Atlantic Drift. As it originates in the Gulf of Mexico, which is one of the warmest parts of the oceans, it is a very warm current, about 25°C. Whilst crossing the Atlantic it cools down gradually, but it still brings much warmth across to the eastern side of the Atlantic. Some of the North Atlantic Drift flows along the Norwegian coast, which is kept ice-free as far east as Murmansk in the Soviet Union. The current is a great source of winter warmth for Britain and helps to create winters which are very mild for the latitude. The Isles of Scilly in Cornwall and the southwest tip of Ireland are able to grow various sub-tropical plants because of the January average temperature of 7°C, which is warmer than along the Mediterranean coast of France at that time of year.

gully A narrow channel on a hillside or slope, formed by the action of water, often because of rapid run-off following heavy rain. **Gully erosion** often referred to as gullying, is the result of rapid run-off on a slope which has no vegetation cover. Small examples of gullying can be seen on some hillsides in Britain. In southern Italy and other Mediterranean lands there are many badly gullied areas, where the vegetation has been removed by man and domestic animals, so that the soil has been washed away by heavy rain. There are many areas with serious gullying in southern Africa and in New Zealand, and most notably in the Badlands in the United States.

H

hachuring see **relief map**

hacienda A large estate, especially in the Spanish speaking countries of South America. Many haciendas have absentee landowners, who leave a manager in charge. They often exploit Indian and mestizo labour. Most haciendas grow a cash product, which may be wheat, cattle or grapes, but generally they also grow food for the workers.

hade The angle of declination of a fault plane from the vertical.

haematite A type of iron ore. It is a rich deposit, often as much as 75% in purity, and always 60% at least. It is frequently found in small lumps, which are referred to as **kidney ore**, because of their shape. Small deposits have been found in Britain in Cumbria, but there are many large deposits in other countries; for example at Kiruna in Sweden, in Western Australia and near Lake Superior in the United States.

hail Small pellets of ice which have been formed by rain drops freezing. When convective currents rise quickly and form large cumulonimbus clouds, the water vapour in these clouds may be carried up to great heights. Large drops of rain may be formed, but if the uplift continues drops may freeze. As they become too heavy to be carried up further, they start to fall, gathering more water as they descend. A further burst of rising air, containing convective currents which are pulsating rather than continuous, will force them to rise again. The extra water may be frozen to form a bigger hailstone. The greater the number of times a stone has to rise, the bigger it will become. In temperate latitudes, hailstones rarely exceed pea size, but in the tropics, where there is a greater amount of convection, stones the size of golf balls sometimes occur. These may weigh up to 0.5 kg, and can do considerable damage when they land.

hanging valley* A valley which is elevated above the level of a main

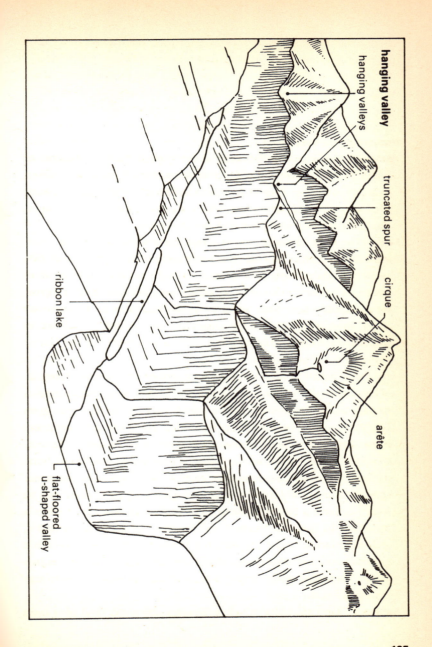

hanging valley

hanging valleys

truncated spur

cirque

arête

ribbon lake

flat-floored
u-shaped valley

105

valley. The main valley will be a U-shaped valley which has been deepened by a glacier. The erosion rate in the tributary valley having been slower than in the main valley, it was not as deeply cut. The tributary valley may be U-shaped if it contained a tributary glacier, or it may be a V-shaped rivercut valley. The stream will flow down steeply from the hanging valley, into the main valley, and it is quite likely that there will be rapids or a waterfall on the tributary stream. Some hanging valleys in Lauterbrunnen in Switzerland and Yosemite in California drop almost vertically for more than 300 m. In England there are impressive hanging valleys in the Langdale and other Lake District valleys.

hardpan A layer of hard material found just below the surface, often in sandstone areas. Rainwater leaches away chemicals including iron oxide and calcium carbonate, and sometimes these solidify a metre or two below the surface. If this happens they form an impervious layer – hardpan – above which there may be marshy patches or puddles, even on permeable rocks. If the layer is mainly iron oxide, it is sometimes referred to as **iron pan**. It occurs very commonly on many of the sandy heathlands in southern England.

hardwood A broadleaved as opposed to a coniferous tree. Hardwood trees are generally much slower growing than conifers and may take 80 to 100 years to reach maturity. British hardwoods include oak, maple, walnut, and cherry. Evergreen oaks and eucalyptus are important hardwoods in the Mediterranean countries and Australia. Moonsoon lands and tropical forests contain teak, mahogany and rosewood.

harmattan A strong northerly or north-easterly wind which blows from the Sahara to Nigeria, Ghana, and other countries on the Guinea coast of West Africa. It is a hot dry wind. If it blows over the humid coastal areas it seems cool and refreshing, and is sometimes called 'the doctor'. However, if can carry dust from the desert, and may be so dry that it shrivels the crops. It is likely to occur throughout the year between 10° and 15° north, but will only reach south to the coast in the winter months.

Harris and Ullman model One of the classical models of urban morphology. See **Ullman, E. R., and Harris, C. D.**.

headward erosion A type of erosion which occurs at the source of a stream. It lowers the land slightly, allowing the stream to rise further back, or further up hill. Over long periods of time, it is possible for streams to erode through watersheds, and this enables them to capture streams flowing in other valleys.

heat island see **urban heat island**

heave see **fault**

heavy industry Any of the industries which produce or use large quantities of bulky raw materials, such as coalmining, steel-making, shipbuilding, and various chemical and engineering concerns. Most of these industries developed in coalmining regions, but many are now located near ports.

hectare A unit of area, equal to 2.47 acres. One hectare equals 10,000 sq m, and 100 hectares are 1 sq km.

Hercynian A period of major mountain formation which occurred at the end of the Carboniferous period and the beginning of the Permian, about 250 million years ago. The name Hercynian is derived from the Harz Mountains in Germany, and Armorican is named after Brittany which was called Armorica in Roman times. In Britain, the Pennines are an example of mountains which were folded and uplifted in Hercynian times. Also called **Armorican**.

hierarchy of settlements A system of grading the various types of human settlement according to size, devised by Christaller in his central place theory.

high A high-pressure system or anticyclone.

high technology The use of highly developed and sophisticated equipment and machinery. It normally involves spending huge sums of money on equipment, but utilizing small amounts of labour. The labour is likely to be highly skilled, trained and educated, and highly paid. Examples of high technology include work with electronics, computers, silicon chips, but also new machinery, such as might be found in very modern automated car-factories or steelworks.

hill farming A system of farming which is generally extensive and pastoral. Sheep and sometimes beef cattle are allowed to roam in the hills and fend for themselves. They have to be rounded up at certain times of the year; for example, for dipping, lambing or selling. The farms are normally located in the valleys, and the animals are often kept near the farms during the winter. A few crops such as hay and oats may be grown as fodder. As soon as the weather improves in the spring the animals are moved up the slopes. A typical hill farm has three distinctive zones: the inbye, the intake, and the fell. The **inbye** is the land in the valley, near to the farmhouse, which may be used for growing crops. The **intake** is the lower part of the valley sides and consists of walled fields. The **fells** are the uplands, which are often unfenced and consist of rough grazing and very poor vegetation. Hill farming is widely practised in the Highlands, in much of Wales, in the Pennines, the Lake District and elsewhere.

hill fog   see under **fog**

hinterland   1 The sphere of influence of a port. It is the area from which exports are obtained, and which receives goods that have been imported through the port. The hinterland of a large port such as London may cover the entire country, but smaller ports, such as Felixstowe or Hull, have a more localized hinterland. The boundaries of hinterlands often overlap, since some goods, such as woollens from Yorkshire, could pass through any of a number of ports; both Liverpool and Hull have handled woollens, and occasionally Yorkshire woollens go through London. A hinterland may be international; for example, Rotterdam handles goods for West Germany, Switzerland, France and Belgium, as well as the Netherlands. 2 The area served by a large inland city, not necessarily a port.

histogram   A graph which shows groups of classes of data by means of a series of columns or rectangles placed side by side. The intervals of values are placed along the horizontal axis, and the height of the column above each value band shows the corresponding frequency. It is a very clear way of summarizing information.

hoar frost   see **frost**

hogsback   A narrow steep-sided ridge which stands up sharply from the surrounding countryside. It may be the result of folding or faulting, followed by differential erosion, in which the softer neighbouring rocks have been eroded more quickly. Examples can be found near Guildford, Surrey, on the Isle of Wight, and elsewhere.

horizon Any of the layers of soil which can be seen in a soil profile. The normal divisions are the A, B and C horizons; A is the topsoil, B the subsoil, and C the bedrock. The A horizon contains fine particles, humus, and decaying vegetable matter. The B horizon contains some fine particles as well as coarser fragments, and contains much more inorganic material than the A. It also contains minerals washed down from the A horizon by leaching. The C horizon is rocky. The boundary lines between these zones are generally rather vague. It is becoming increasingly common to divide the C horizon into C and D. In this case the C is weathered rock fragments, and D is the real bedrock.

horn A pyramid-shaped peak such as the Matterhorn in the Alps. The pyramid would have been steepened and carved by glacial erosion, the ice action and freeze-thaw activity having formed corries, which wore away the sides of the pyramid. The summit becomes smaller and steeper to form a peak. It is often possible to reach the summits of horns by ascending along the line of arêtes between the corries. Everest is a horn, and Snowdon has been eroded on three of four sides to form a near pyramid peak. Also called **pyramid peak**.

horse latitudes Either of two areas of high pressure found near 25° north and south of the equator. The air rises at or near the equator because of low pressure and then circles in the atmosphere, but much falls near the tropics of Cancer and Capricorn. The falling air gives rise to fairly permanent regions of high pressure, the horse latitudes. Both the trade winds, which blow from the horse latitudes towards the equator, and the westerly winds, which blow from the horse latitudes towards the poles, are caused by the high pressure. It is suggested that the horse latitudes were so called because in the days of sailing vessels it was easy to become becalmed in these areas, in which case everything possible would be thrown overboard in order to lighten the vessel. Apparently it was not unknown for the horses to be discarded at this time.

horseshoe lake see **oxbow**

horst see **block mountain**

horticulture The growing of vegetables, flowers and fruit. It is generally carried out on small farms or market gardens, and is very intensive; there is much horticulture in greenhouses, too. Horticultural concerns are normally located near towns, in order to get the produce to market quickly; for example, in southern Essex, near London, and just outside New York and Washington in the eastern United States. There are also some favoured areas where early vegetables can be grown because of climatic advantages. In Britain, southwest England and parts of Pembroke can grow early potatoes and some flowers; the Scilly Isles grow daffodils which are picked in January, and the Channel Islands are able to grow tomatoes and flowers which are ready earlier than mainland products. In the United States, Florida grows early fruit and vegetables, which are sent north to New York, and California also grows early products. Also called **market gardening, truck farming** (US).

hosiery Socks, stockings, tights, etc. Traditionally the hosiery industry has been important in Leicestershire and Northamptonshire.

hotspot An area on the earth's surface where the crust is quite thin and volcanic activity can sometimes occur, even though it is not at a plate margin. The Hawaiian Islands are at a hotspot in the north Pacific.

Hoyt, H., U.S. researcher who developed the sector model of urban growth. In 1939 he wrote about his views of urban growth, based on his studies of land use patterns in 142 American cities. His ideas differed from those of Burgess, who regarded concentric growth as the most important, whereas Hoyt saw that sectoral variations could be related to the development of transport links. His is one of the three classical models of urban morphology.

human geography A major section of geography which includes the study of people, work, life-styles, and settlements. It also considers how man influences and changes the environment, as well as being influenced by it.

humidity The amount of water vapour in the atmosphere. See also **absolute humidity, relative humidity**.

humus The dark organic material in soils, produced from plant and animal matter. As plants and animals die and decay, their decomposing remains in the soil form humus. It will be ploughed up in cultivated land, and possibly added to with manure and fertilizer. In areas where there is no cultivation, it gradually decays and is washed downwards by rain. Some of it is then lost by leaching, but much will be reused in the next generation of plants. The humus layer varies from soil to soil. It is often black in some of the richer soils, but even in the best soils it can be used up by repeated ploughing, especially if monoculture is practised.

hunter and gatherer A member of a tribe or group of people, especially in areas of tropical forest, who survive by hunting animals and birds, and gathering wild fruits and berries. They will also fish to add to their food supplies. They have to move around from place to place in the search for food, and so they do not build permanent homes. Lean-to shelters of leaves and branches are often adequate in the warm climatic conditions of the tropics. The Pygmies are an example of this way of life, and other groups are found in Africa, South America and southeast Asia. Only small numbers of people still pursue this way of life.

hurricane 1 A tropical cyclone in the Caribbean or Central America. 2 A hurricane-force wind; that is force 12 on the Beaufort scale.

hybridization The crossbreeding of plants to produce new varieties, which are developed for special qualities, such as resistance to disease or pests, or the ability to ripen more quickly or give a greater yield. Hybrid varieties have been produced for many centuries, and even the earliest cultivated plants were selected from natural crossbreeding. Research stations, such as the International Rice Research Institute in the Philippines, continue to produce new varieties each year.

hydraulic action The erosive effect of water. The force of moving water can be an effective agent of erosion, rather like a battering ram. It is important both in rivers and on coastlines. In rivers the movement of currents causes hydraulic action, though if a load of silt or sand were added the process would be called corrasion. Along the coast, breaking waves sometimes trap water in cracks and joints and this adds to the hydraulic effect.

hydro-electric power Electricity produced from falling water. Natural or artificially created waterfalls provide the head of water to drive turbines to generate electricity. If there are no waterfalls, dams are built, and the water drops from near the top of the dam down to a power station at the foot. Most hydro-electric power stations are located in highland areas where the climate is quite wet; for example, the Alps, Norway, and

Sweden. In recent years, because of improved technology, it has become possible to harness the power of large slow-flowing rivers, such as the Angara in Siberia, and the Amazon in Brazil. In Britain there are numerous hydro-electric schemes in Scotland, several in Wales and Ireland, but none in England. At Cruachan in Scotland and, more recently, at Dinorwic in Snowdonia, pumped storage schemes have been developed. The electricity produced during the night, when there are not as many customers, is used to pump water back into the lake or reservoir, so that it can be used again to produce electricity.

hydrological cycle* The water cycle. Water is evaporated from the sea; some falls back into the sea as rain, but much is carried over the land. There it falls as precipitation, and by surface run-off or infiltration and seepage it gradually moves back into the sea. Less than 1% of the world's water is involved in this cycle at any particular time; 97% of all water is in the sea, and much of the remainder is snow or ice.

hydrology The study of water on or under the earth's surface, and its uses, availability, conservation, etc.

hydroponics The cultivation of plants in nutrient solutions rather than in soil. Soil provides plants with nutrients and a medium in which they can stand up. If those two requirements are met in other ways, soil becomes irrelevant. Nutrients can be provided in the water, and trays of gravel, sand, stones, or even polystyrene make a suitable growing medium. Hydroponic techniques are very useful in places where there is a shortage of soil, such as some of the semi-arid parts of Africa. Such areas have long hours of sunshine, which is very suitable for growing plants, providing they are watered and fed artificially. Hydroponic methods are also used in the arctic areas of Canada and the Soviet Union, where long hours of sunlight in summer enable fresh vegetables to be grown. They are important in the production of food in several Middle Eastern countries and also in Singapore, where there is a great shortage of land. Some crops are grown hydroponically in Britain, in sheds where an artificial climate can be created to overcome the vagaries of the British weather.

hygrometer An instrument for recording the relative humidity of the air. It consists of two thermometers, one a dry bulb and the other a wet

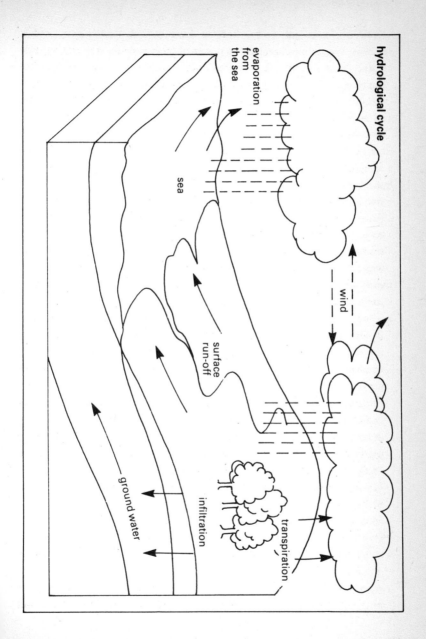

hydrological cycle

evaporation from the sea

sea

wind

surface run-off

transpiration

infiltration

ground water

113

bulb. The wet bulb is covered by a small piece of muslin cloth, which is kept damp by water from a small reservoir at the end of the hygrometer. Evaporation from the cloth reduces the temperature of the wet bulb, and the difference in temperature between the wet bulb and the dry bulb shows the relative humidity. Special tables are available to give the correct humidity. The two thermometers can be kept in a Stevenson screen, or can be placed on a mount somewhat like a football rattle, which can be whirled round. This type of hygrometer can be used out in the field to give approximate shade temperatures in an open and exposed location.

hypabyssal A term denoting igneous rocks which formed neither at the surface, nor at great depth. They sometimes contain a few large crystals as well as many small crystals, and are intermediate in grain size between plutonic rocks, which formed at depth, and volcanic rocks, which formed on the surface. Hypabyssal rocks are often found in sills and dykes.

hypermarket A very large store or shopping complex selling a wide range of necessities and consumer goods. Normally located on the outskirts of towns, hypermarkets include a large supermarket and various other types of shop, as well as ample parking space, possibly a garage, a children's play area and a bank. They developed first in the United States because of the wealth which enabled most families to become car owners. France was the first European country to develop similar out-of-town shopping complexes, but the idea has now reached Britain and many other countries.

hypothesis An idea or a suggestion or a theory which will be proved or disproved by studying the relevant facts. A problem solving technique will be necessary to investigate the hypothesis.

I

ice Frozen water, or snow which has passed through the firn stage to form glacier ice. The relatively low density of ice means that it floats.

ice age A prolonged period of colder climatic conditions, during which snow and ice covered large areas of the earth. There have been several ice ages in the past. The most recent began about 2 million years ago, and is often referred to as the Ice Age. During this period, ice advanced for several thousand years, and then retreated for about the same period. An advance of ice is called a **glacial phase**, and the retreat is an **interglacial**. We are living in an interglacial at present, although there are still ice-age conditions over Greenland and Antarctica. During the glacial phases ice covered Britain as far south as the Bristol Channel and the Thames estuary. Areas further south experienced permafrost and tundra conditions. When the ice melted, finally disappearing about 11,000 years ago, the climate warmed up, enabling different types of vegetation to colonize southern England, and eventually spread further north.

iceberg A floating mass of ice which originated on land as part of a glacier or icesheet. As the ice moves slowly outwards from a snow cap, some reaches the sea. It floats and is gradually broken off by the movement of waves and tides. Some icebergs may be several hundred metres in size, but they gradually melt and shrink as they float across the oceans. Icebergs from Greenland are often quite tall, whereas those from the Antarctic are flat and tabular in shape. Up to 80 or 90% of an iceberg is below the surface of the water. As they melt, icebergs sometimes turn over; this readjusts the balance. One of the major areas for icebergs is on the route of the cold Labrador current, which brings them southwards along the coast of Greenland, towards Newfoundland.

icecap A large expanse of snow and ice covering a mountain range, an island, or few square kilometres of land, as for example in parts of

Iceland, some of the islands off the northern coast of Canada, and Spitzbergen. An icecap is a smaller version of an icesheet.

icefall A section of a glacier where there is an abrupt change of gradient and the ice goes down a steep slope. As it does so, many transverse crevasses form and at the bottom of the icefall, there will be a mass of séracs where the ice has been compressed again and the crevasses have closed up. An icefall is a very difficult area to cross and major icefalls in the Himalayas and Antarctica always create problems for explorers and climbers. Great Khumbu near Everest is a famous example, and there are many small examples in the Alps.

icesheet A large expanse of snow and ice which covers a land mass; for example, Antarctica. During the Ice Age an icesheet covered much of Norway and Sweden, and spread across the North Sea to join the ice of Britain. It also covered parts of northern Germany and Denmark. At the same time, there was another icesheet across Canada and northern United States. Icesheets can cover entire mountain ranges, and can spread across lowlands and oceans. The ice may become very thick, and in Greenland there are places where the ice measures 3000 m. Antarctica has areas of similar thickness. If all the ice melted from these two icesheets, the level of the oceans could rise by as much as 100 m.

iceshelf A sheet of floating ice on the sea. There are large areas of shelf ice round Antarctica, and small areas on the edges of the Arctic; for example, off the northern coast of Canada.

Igneous A term denoting rocks of volcanic origin which were formed from magma. If the magma solidified beneath the surface, cooling would have been slow; if it solidified at the surface, cooling would have been more rapid. The faster the cooling process, the smaller the crystals which make up the rock. All igneous rocks are crystalline. In addition to grain size, the other important differences found in igneous rocks are related to the chemical content of the minerals in the rock. Acidic minerals, including quartz, some felspars, and some mica, are generally light in colour. Conversely, basic minerals are generally dark in colour; they include some felspars, biotite mica, hornblende and augite. Igneous rocks formed at depth are called plutonic, those formed at the surface are called volcanic and those formed between are hypabyssal.

Volcanic rocks are extrusive, and the commonest variety is basalt. Hypabyssal rocks are found in sills and dykes, and they are intrusive rocks. Plutonic rocks form in large bodies of magma called batholiths, stocks, and bosses. The commonest plutonic rock is granite, which is acidic in mineral content; the basic equivalent is called gabbro. Igneous rocks do not have bedding planes, and they do not contain fossils. They are mostly quite resistant to erosion, and often form high ground; for example Dartmoor, Bodmin, the Malvern Hills, parts of the Lake District and the Cairngorms.

illuviation A process of deposition in a soil. This normally takes place in the B horizon; materials removed by eluviation from the A horizon are deposited in the illuvial layer. It is possible for the particles to form a layer of hardpan.

immediate run-off see **surface run-off**

immigrant A settler from another country or region. Immigration is generally used to describe international movements only.

impermeable A term denoting rocks which are non-porous and therefore do not absorb water. Granite is a good example of an impermeable rock; however, it is pervious, because water can pass through its joints. An exception is clay, which is porous. It contains numerous tiny pores, but when these have filled with water they prevent the movement of water through the rock, making it impermeable.

impervious A term denoting a rock which is not pervious. It will not allow the passage of water because it contains no joints, cracks, or fissures; for example, slate and shale.

import A commodity or article brought into a country from abroad. Imports may include minerals to be purified and processed, sources of fuel and energy, manufactured goods or foodstuffs. Imports generally have to be paid for out of the income from exports. Many Third World countries export raw materials, but have to import most of their manufactured goods.

inbye see under **hill farming**

incised meander An old meander which has become deepened and has cut down into the landscape. This is generally the effect of extra erosion caused by rejuvenation of the river, sometimes as a result of river capture providing extra water, or of uplift of land near the source of the river producing an increased gradient. The extra erosion causes the river to flow at a level below the former flood plain. Incised meanders may be ingrown or entrenched. In an **ingrown meander** some lateral as well as vertical erosion has occurred, and the cross-section of the valley may be asymmetrical because of lateral erosion on the outside of the bend only. Entrenched meanders have been affected simply by vertical erosion, and the valley cross-section is even and symmetrical.

industrial crop   A farm product which is grown in order to provide the raw materials for an industry. For example, jute grown in Bangladesh supports a sacking, linoleum and carpet industry. Cotton, hemp, flax, and rubber are industrial crops, and potatoes are sometimes used to make industrial alcohol. Sugar beet and sugar cane can both support industries, and wool from sheep is another important source of industrial raw material.

industrial estate   A specially selected site, generally on the outskirts of a town, where various industries are located in new factories. The estate is generally purpose-built and will be well equipped with road transport, electricity, water supply, etc. Industrial estates generally contain light industries only. Many have been set up in the old declining industrial areas, such as south Wales and northeast England, in order to provide new industries and employment to make up for the closure of coal mines and steelworks. There are industrial estates in new towns such as Crawley, Bracknell, Milton Keynes, and most other towns have one or more industrial estates nowadays.

industrial inertia   see **geographical inertia**

industrial location   Old heavy industries such as steel manufacture were always located on coalfields, especially if iron ore and limestone was available locally, as in south Wales and Middlesbrough. Most new steelworks are now located on the coast, so that materials can be transported by sea. The provision of transport has become a more important factor than proximity to raw materials. In the case of some mineral purification, the industry is located near the mine; for example, copper and tin mining in Bolivia. In other cases a mineral may be transported many miles for refining; oil from the Middle East is refined in Japan and Europe. Aluminium refineries are often located near sources of hydro-electricity, because the power is more significant than the raw material, as at Kitimat in western Canada and Fort William in Scotland. Some industries are now located near their markets because they need only small quantities of raw materials, and labour and markets have become more important than raw materials as a locating factor; for example, electrical and electronic industries near London. Sometimes locations may be chosen because of political decisions. Governments want to create jobs in areas of high unemployment, and so they give tax incentives, and grants to companies which will locate their factory in a place selected by the government. An example of this is the Nissan factory near Sunderland.

Industrial Revolution The period when the development of steam power and many other inventions enables a great deal of rapid progress to be made in a variety of industries. This mainly took place in the years 1780 to 1820. As steam power developed and industries expanded, more workers were needed for more factories. This contributed to a rapid growth of wealth and of the urban population. The early industries were located on coalfields; for example in south Wales, Yorkshire, Lancashire, northeast England, and central Scotland. Steel, engineering, chemicals, cotton and woollens were the major developments. After the Industrial Revolution, there was continuous change in the industrial areas, as newer and more expansive developments took place. However, the area around Ironbridge and Coalbrookdale in Shropshire never really developed; having been one of the most important steel areas in the world at the time of the Industrial Revolution, the area had declined by the end of the 19th century. The area, now part of Telford new town, was a decaying and derelict area until the 1960s, when the Ironbridge Museum Trust and Telford Council decided to restore many of the buildings and create museums to show the area as it must have looked during the Industrial Revolution.

infiltration The downward movement of water from the surface into the soil. The rate of infiltration depends on the amounts of rainfall and its intensity, the vegetation cover, the compactness of the surface, the porosity and permeability of the rock, and the amount of water already in the soil.

Informal economy Casual employment often with irregular hours, using children and sometimes of dubious legality. This is found in many parts of the Third World, where there are many street traders. Small industries are increasing in many cities, as well as services such as shoe-cleaning. It is believed that the informal sector employs at least 50% of the workers in some large cities. There is also an informal economy in advanced countries as workers do part time jobs whilst receiving state assistance, and some people work at two jobs, by doing evening or night shifts in addition to their normal working hours.

infrastructure A basic framework of roads, power and water supplies, schools, hospitals, etc.

ingrown meander see **incised meander**

inlier An area of old rocks surrounded by younger rocks. It sometimes consists of quite resistant rock, which has been slow to erode, as for example, the Malvern Hills. Sometimes an inlier is the result of erosion of an anticlinal structure; the exposed underlying rocks are surrounded by the younger rocks of the structure, as in the Vale of Pewsey.

inner city A residential area in or near the centre of a city. Many inner city areas suffer from old and decaying housing, often in need of repair or renewal. Gradually they are being redeveloped, but there are frequently cases of vandalism and other social problems, and consider-able expenditure is needed to improve conditions in the inner cities, since land prices are very high. Major developments have taken place in some locations, such as Tower Hamlets and parts of the old dockland areas of London.

input Any of the items (energy, raw materials, etc.) which are put in to a system. For example, in farming, seed, fertilizer and sunshine would be inputs, and in a steelworks the major inputs would be iron, coal and limestone. All systems have inputs and outputs.

insolation The amount of energy received from the sun. The amount varies according to the length of day, latitude and condition of the atmosphere. Polar regions receive long hours of insolation in summer, but none at all in the winter, while equatorial regions receive similar amounts throughout the year. Deserts have clear skies and receive more insolation than some equatorial zones where cloud is present for a few hours each day. Pollution, as well as cloud, can restrict the amounts of insolation reaching the earth. Before the clean air acts of the 1950s, London suffered from haze and smogs and received far less sunlight than it does nowadays.

instability The condition of an air mass with a lapse rate which is greater than the dry adiabatic lapse rate (average = 1°C per 100 m). If it rises, the air mass is likely to continue rising to such a height that condensation may occur. Therefore instability of air means that rain is a possibility. When unstable air masses are passing overhead, cloud will form in many places, especially over mountains or any areas where there has been extra heating. This means that instability is greater during daylight hours, when more heating is likely to occur.

intake see under **hill farming**

integration The process of carrying out all the distinct stages of an industry in one location. Large new steelworks are generally integrated; raw materials go in at one end of a large factory, and finished products come out at the other. Many car factories are integrated. A conveyor belt moves from one end of the factory to the other, and completed cars can be driven away at the end of the belt. Integration has also taken place in the cotton and woollen industries; the raw materials are cleaned, spun, woven and made into articles of clothing or other items, all within one factory.

intensive farming Any system of farming which gives a high return per hectare. Generally much labour, or capital is necessary in order to produce high yields, and the land is cultivated annually and not left fallow. Most parts of Britain and many parts of Europe are intensively farmed. Market gardening is one of the most intensive land uses, and greenhouse cultivation is especially intensive. Frequent, possibly daily, inputs are vital for the commercial and economic success of some intensive farms. A few areas of farmland are intensive yet subsistence, and this applies particularly to some of the rice growing areas of south-east Asia, such as the Si delta in southern China, and the lowland areas in Java.

interglacial see **ice age**

interlocking spurs A series of spurs on alternate sides of a river valley.

They are the remnants of high ground which has largely been eroded by the river. The meandering course of a river causes erosion to take place on alternate sides of the valley. In places where there has been no recent erosion the higher ground will protrude as spurs. From the valley floor it is possible to see the spurs jutting out and overlapping, or interlocking. This can be seen clearly on a small scale in the upper course of a river, but it also occurs in the middle course. By the time the lower course has been reached, the valley width will be too great for the spurs to overlap, and the valley walls are normally so reduced in height that any remaining spurs are quite low.

intermediate technology see **appropriate technology**

intermittent stream A stream which flows for only part of the year. Intermittent streams dry up during the dry season, which is the winter in savanna regions, and the summer in Mediterranean regions. In Britain some streams called winterbournes dry up when the water-table becomes lower during the summer. Desert areas often have intermittent streams which flow for a few days or weeks after heavy rainstorms. See also **exotic stream, perennial stream**.

intermontane Between mountains. The term is generally used to describe a plateau or basin. An intermontane basin is a large hollow situated between two mountainous areas; for example, the Great Basin of Nevada, Great Salt Lake. An intermontane plateau is a high-level area which is lower than the surrounding land. The plateau of Tibet is much lower than the adjacent Himalayas, and in South America the plateaus of Peru, Ecuador and Bolivia reach over 3000 m, but are dwarfed by the Andean ranges alongside them.

international date line A theoretical line which approximates to the 180° line of longitude. It is used to overcome a time problem caused by travel; anyone crossing the line from west to east has to move the clock forward by 24 hours, and jump one day, whereas anyone going from east to west puts the clock back, and has to live the same day twice. The changes of time are caused by the earth's rotation. As the earth spins round, different parts turn to face the sun and receive daylight. Travelling east from Greenwich, the time becomes later, at the rate of one hour per 15° longitude. Moving west from Greenwich, times are behind Greenwich Mean Time, by one hour per 15°, so that 15° west will be two hours behind 15° east, 30° west will be four hours behind 30° east and so on. Eventually, there is a difference of 24 hours between 180° west and 180° east, but since the two lines of longitude are the same the international date line was devised to solve the problem.

intertropical convergence zone   see under **doldrums**

intervisibility   The visibility of one point on the landscape from another. It is possible to use ordnance survey maps and work out whether from one particular location it is possible to see other locations in the surrounding countryside. In order to determine whether hills, buildings, or woods would block the view, it is necessary to draw a careful cross-section.

intrazonal soil   A soil which develops in a particular environment, irrespective of climatic conditions. Examples include rendzinas, which develop on limestones, and gley soils, which develop on waterlogged land.

intrusion   The process by which intrusive rocks are formed.

inversion layer   see **temperature inversion**

inverted relief*   A type of relief in which former uplands have been so eroded that they now lie below the surrounding areas which were previously lower. When folding takes place, the upfolded rocks in the anticline are often stretched and cracked, which means that they are more easily eroded than the more compressed and consolidated rocks in the syncline. Erosion over a few thousand years can wear the anticlinal area down to a lower level than the syncline. This type of relief is surprisingly common. In the Vale of Pewsey a large anticline has been beheaded to expose the underlying rocks. The part of the landscape which was formerly the highest has become the lowest. The same is also true of the Weald, where the highest land, formerly in the middle, is now located along the chalklands of the North and South Downs. Similarly, Snowdon, which was formerly part of a synclinal structure, is now the highest mountain in Wales.

iron and steel   A heavy industry which for a long time has been regarded as a vital part of any nation's industrial development. During the past 20 to 30 years Japan, Taiwan, South Korea, and many other countries

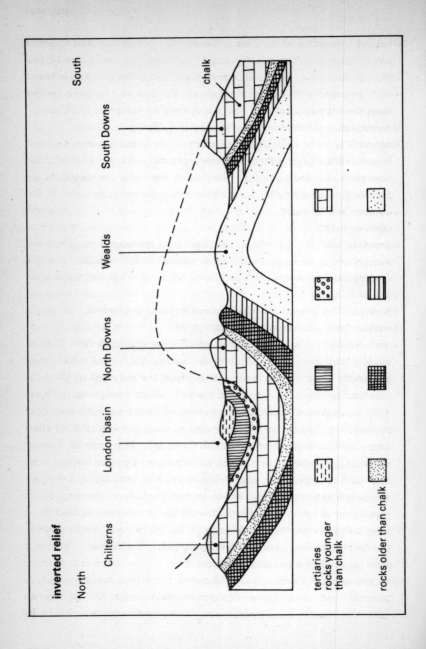

inverted relief

North   South

Chilterns

London basin

North Downs

Wealds

South Downs

chalk

tertiaries
rocks younger
than chalk

rocks older than chalk

have developed steel industries, and so the European steel industries have declined because of the growth of competition. In Britain the location of steelworks has changed. They used to be on the coalfields, such as south Wales and central Scotland, especially in places where there were small deposits of iron ore as well. Nowadays they are mostly on the coast, so that large quantities of iron ore can be brought in from overseas, from countries such as Australia, Brazil and Canada. The new steelworks are large integrated factories, which need a big area of flat land, and reclaimed land near the coast has been found very suitable, for example, at Port Talbot and Llanwern in south Wales.

iron pan   see **hardpan**

irrigation   The supplying of water by artificial means to an area where there is a shortage, usually for the purpose of growing crops. There are various methods employed. Simple old-fashioned methods like the Archimedes' screw and the shaduf are still useful for lifting water from one level to another, and water wheels are used for the same purpose. Wells are increasingly numerous in many parts of the world, and simple wells, which are merely holes in the ground, are being replaced by tube wells which are lined and can reach deeper supplies of water. There is often a surfaced area round the top of the well in order to reduce erosion by humans and animals. Most important nowadays are the large schemes for irrigation in which water is available throughout the year. This is called **perennial irrigation**, and requires a dam and a large lake in which vast quantities of water can be stored. Some of the largest lakes, such as Lake Nasser behind the Aswan dam, contain enough water to last for two years, even if there were no more rainfall. The dams often supply hydro-electric power as well. Leading from the lake will be a network of canals, which not only provide reliable supplies of water throughout the year, but also enable a greater area of land to be irrigated. Some of the major irrigated areas are in deserts; for example, the Nile valley in Egypt and the Indus valley in Pakistan. There is a great deal of irrigation in regions with a Mediterranean climate, including southern France and California. Even quite wet areas such as parts of India and China, can benefit from irrigation in order to guarantee reliable supplies of water, or to enable a second crop to be grown in the drier winter months, when temperatures are still high enough for rice and other plants to grow successfully.

isobar A line on a weather map joining places of equal pressure. The pressure readings are normally taken at or near sea level, although upper atmosphere pressure readings are used to help with weather forecasting and also in connection with planning routes for aircraft. Isobars drawn on synoptic charts indicate weather patterns, including the location of depressions and anticyclones. Average isobars can be drawn on continental or world maps to show the generalized pressure distribution. On daily weather maps the isobars are generally drawn at intervals of 4 mb.

isohyet A line on a weather map joining places with equal rainfall. Isohyets are generally drawn to show the distribution of rainfall over a year, or perhaps for the summer months, but they could be drawn for any length of time.

isoline A line on a map which is joining places having the same value of any selected element. An isobar or isohyet is a type of isoline, and so too is a contour. Also called **isopleth**.

isotherm A line on a weather map joining places of equal temperature. Before maps are drawn, temperatures are normally reduced to their sea level equivalent. They can be drawn to show the average January or July temperatures, or to show the termperature at a given time on any particular day. Thus they can show precise temperature information, or give averages over a longer period of time.

isostasy The state of balance or equilibrium of the earth's land masses. The continental rocks which are made up of sial, sit on top of the denser rocks, the sima, which make up the ocean floors. When the continents are large they weigh down into the sima, but as erosion reduces their weight, they gradually rise. This is isostatic readjustment. During ice ages, the weight of the ice presses the continents down. As soon as some ice melts, the land masses begin to rise because of isostasy. When this happens the former coastlines are elevated to form raised beaches, such as those which can be seen in many parts of Scotland and along the Scandinavian coastline.

J

jet stream A strong current of air blowing through the atmosphere at a high altitude. The main jet streams are in middle and sub-tropical latitudes. The temperate jet stream flows from west to east, though along a wave-like route, going north and south as it winds all round the world. It can blow at speeds of up to 370 km/h at times, and so is an important factor for aircraft. If planes can fly with the jet stream, which often blows at 10,000 to 13,000 m, it can increase the speed of the journey and save fuel costs. If the aircraft is going westwards across the Atlantic, it will try to fly above or below the jet stream. Jet streams influence the weather systems in the lower layers of the atmosphere, as depressions tend to follow their route. Jet streams change position fairly slowly, and once established in a particular location may persist for a week or two. If the line of the jet stream is passing over Britain, a series of depressions can be expected. If on the other hand the jet stream is to the north or to the south of Britain, some settled anticyclonic weather will be experienced.

joint A vertical or near-vertical crack in a rock. Joints occur in sedimentary rocks because of shrinkage or weathering along a line of weakness. In igneous rocks they are formed as a result of cooling and contraction, as magma solidifies. Joints are much smaller than faults and there is little or no movement on opposite sides of the joint; they are often a site of weathering and erosion. Cliffs and steep slopes may be influenced by the positioning of joints. Joints are very frequent in carboniferous limestone and also in granite and basalt. Basalt often cools into hexagonal shaped blocks because of the vertical cracks which occur during cooling.

joule A very small unit of energy. One joule equals 0.239 calories; one megajoule (MJ) equals 239 calories; and one gigajoule (GJ) equals 239,000 calories. About 10.5 MJ, or 2500 calories are needed as the daily food requirements of an average working person. The amount of solar radiation can also be measured in joules. Aberystwyth in central Wales receives 210 joules per sq cm per day in the winter, but 1880 per day in summer. Brisbane in Australia varies from 1150 to 2480 and Algiers from 880 to 2800.

jungle A type of tropical or monsoon forest with dense undergrowth. Many tropical forests, which have rain all through the year, have so much growth of vegetation that there is a dense canopy through which sunlight cannot penetrate. In these areas near the equator there is no undergrowth, and no jungle. It is where there is a short dry season and the forest is less dense and more open that undergrowth flourishes. The most impenetrable areas for jungle growth are in the monsoon areas of southeast Asia, though even here large areas of woodland are not jungle.

Jurassic A geological time period which occurred about 150 million years ago. Britain was covered by warm tropical seas at this time and many sediments were laid down. Jurassic rocks are clays, sandstones and limestones. The best-known limestone is the oolitic limestone of the Cotswolds, which is often used as a building stone, as in Bath. Many fossils are found in Jurassic rocks, including ammonites, belemnites and some dinosaurs.

K

Kainozoic see **Cainozoic**

kame A fluvio-glacial deposit formed by streams flowing on to or beneath ice. Streams flowing down mountain sides onto glaciers deposit sand and gravel at the point where the stream first reaches the ice, or where the water goes down into a crevasse. The accumulations of sand and gravel, gradually fall down to the valley floor, as the ice melts. Some kames are the result of deltaic formations created by sub-glacial streams. Kames are generally small features, 20 to 30 metres in length or breadth, though occasionally kame terraces form along the side of the ice and they may be elongated. Examples can be found on the southern slopes of the North Yorkshire Moors, and in northern Norfolk.

kanat see **qanat**

kaolin see **china clay**

karez see **qanat**

karst scenery The type of landscape found on limestone where most of the water has gone underground. Karst takes its name from the Karst region of Yugoslavia. Solution weathering wears away much of the limestone, and rivers also erode the rock. Most rivers are underground, and there are caves and large caverns. The largest caverns may collapse to form gorges, and gradually the entire area of limestone may be worn away. The surface features of the limestone usually include clints and grykes, limestone pavements and swallow-holes. There are good examples near Malham and Ingleborough in Yorkshire, and the Causses region of France have many kilometres of karst scenery.

katabatic A term denoting a wind which blows downhill. Katabatic winds are most likely to occur at night when air which has been cooled by radiation accumulates in the valleys and begins to flow downhill. They can also be found near glaciers and icecaps when the cold air from above the ice starts to move downhill to lower levels. Winds of this type may blow during the day as well as the night; for example, in Greenland. The opposite of katabatic is **anabatic**. Winds which are anabatic blow uphill, because of heating during the day. They are less powerful and less frequent than katabatic winds.

kettle hole   A small hollow 10 to 20 m across, situated in morainic debris or on an outwash plain. A kettle hole is formed when a block of ice dumped with glacial deposits, slowly melts; as it does so, the remaining debris collapses to form a hollow. Kettle holes are often marshy, and may even contain a small pond.

khamsin   A hot southerly wind which blows across Egypt towards the Mediterranean. It is very dry, with relative humidity as low as 10% on occasions. This will crack the skin, as well as woodwork. It blows into the front end of a depression which is passing along the Mediterranean. Large quantities of dust are sometimes carried by the khamsin, which is a similar wind to the sirocco of Algeria.

kibbutz   A type of cooperative farm and settlement found in Israel. All the members of the kibbutz help with the work, whether farming, cooking, or cleaning. Many kibbutzim have been created in southern Israel on land reclaimed from the desert by means of irrigation.

kidney ore   see **haematite**

knick point   Also **nick point**   A break in the long profile of a river which may be as a result of rejuvenation caused by uplift of the land. Alternatively, it can result from river capture. It is normally marked by a small fall or some turbulent water, but in time the knick point will be smoothed off by river erosion.

knoll   A small rounded hill.

knot   A measure of speed, sometimes used instead of miles per hour. It is slightly faster than miles per hour, as it represents one nautical mile per hour. It is a relic from the days of sailing ships.

kolkhoz   (*Russian*) A collective farm.

L

labour intensive A term used to denote an industry or business activity which uses proportionately more labour than capital or machinery. This can apply to agriculture as well as industry. In some areas, most of the farm work is done by hand, as for example many of the rice growing areas of Asia. Some factories still employ a large labour force; for example, some sweet manufacturers or toy makers, but many large factories are now capital intensive and use machinery and automation. Labour intensive methods are probably more appropriate in the less developed countries, as they have large supplies of labour available but not much money to buy machinery. See also **capital intensive**.

laccolith A igneous intrusion which has forced its way between the existing strata and is therefore concordant with the structure. Often the overlying rocks bulge upwards, forming a small hill on the surface. Laccoliths are formed of magma; they are smaller than batholiths, and generally have a flattened base and a domed upper margin.

lacustrine Relating to a lake. Lacustrine deposits are those which have been laid down on the bed of a lake. When the lake is filled up and disappears, which happens to all lakes eventually, a flat and fertile lowland will remain. The Red River valley south of Winnipeg consists of lacustrine deposits from a post-glacial lake, Lake Agassiz. The Vale of Pickering in Yorkshire was formerly a lake, and many of the valleys in the Lake District, such as Langdale, formerly contained lakes.

lagoon A stretch of shallow water partially or completely cut off from the sea. If it is completely cut off, there will be some percolation of water through the ridge or spit which separates the lagoon from the sea. Lagoons are often found behind spits, bars and coral reefs.

land breeze A light wind blowing seawards from the adjoining land, usually at night. During anticyclonic conditions, when local weather controls develop, the land cools down more quickly than the sea at night. Air starts to rise from the relatively low pressure of the warm sea, and cool air drifts out from land to sea, until the sun has warmed up the land in the morning. During the day a reverse flow, or sea breeze, will develop. Land and sea breezes both develop because land heats up and cools down more rapidly than sea. In some tropical areas, where land and sea breezes are more frequent and more powerful than in Britain, fishermen can go out to sea with the land breeze very early in the morning; by mid-afternoon a sea breeze will have developed to blow them back to shore.

landform A landscape feature, such as a hill, plateau, or valley. Geomorphology is the study of landforms.

land reform The redistribution of land. A common land reform is the breaking up of large estates and redistribution of the land among the peasants who were previously landless labourers. Land reform endeavours to reduce inequalities, as in many countries most of the land is owned by a few people. There have been revolutions in several countries because of this unequal distribution. In many cases the land reform has helped small farmers to grow more food and also earn money with cash crops. In other places the new landowners have not really known what to do with the extra land. Help and advice is often an essential part of any land reform programme. A totally different type of land reform has been taking place in parts of Africa, in Cyprus, Germany, and other countries. Here there are many farms which are fragmented, each farm consisting of several tiny plots of land. Attempts have been made to consolidate the farms and arrange for all parts of the farm to be together, rather than dotted about over several miles of countryside.

landscape The scenery, including the geomorphological features as well as the man-made features, such as fields, buildings, roads and cities.

landslide A usually rapid movement of rocks, soil and vegetation down a slope. It may be caused by an earthquake, but is generally the result of rain soaking the ground; the effects are exaggerated if there is an impervious layer beneath the surface, as for example on the cliffs near Folkestone, and the Dorset coast near Lyme Regis. In colder climatic areas the saturated rock may be affected by freezing and thawing of the water, and this increases the likelihood of landslides. Also called **landslip**.

land tenure   The system of land ownership or rental. Owner-occupiers are common in Britain, but most farmers are tenants. There are also sharecroppers in many parts of the world.

land use   The ways in which land is used can be plotted on maps. Land utilization surveys have been conducted over the whole of England, as well as in other parts of the world. The present land utilization survey produces maps on a scale of 1:25,000. Land use maps can also be drawn for urban areas; they can reveal zones of industrial use, business use, expensive housing, etc., and on large-scale maps it is possible to show the use of each building.

land value gradient   The scale on which the average value of land declines in proportion to its distance from a city centre. There are exceptions to this generalization, and a line graph can be drawn to show the variations in land value. There is a general decline, but small peaks will appear at important crossroad areas and near suburban shopping centres.

lapilli   see **pyroclastic rock**

lapse rate   The rate at which temperature decreases with height above the ground. The average lapse is about O.6°C per 100 m, and this is called the **environmental lapse rate**. The rate will vary according to the water vapour content of the atmosphere. If dry air is rising, it will cool more quickly, at the dry adiabatic lapse rate (about 1°C per 100 m). If the air is wet the cooling rate will be slower, at the saturated adiabatic lapse rate (about 0.5°C, on average). The reason for the lower rate of decrease is that the water vapour gives off latent heat as it condenses, and this extra heat slows down the temperature change.

lateral erosion   Sideways erosion. This occurs in rivers on the outside of meanders, and gradually the valley becomes wider. Lateral erosion eats into bluffs and slowly wears them away. Corrasion is the main process involved, but there may also be some hydraulic action and corrosion.

lateral fault   see **tear fault**

lateral moraine   A linear pile of rocks deposited by a glacier. As the glacier moved, rock fragments were pushed out to the sides, where they were left when ice melted. In a U-shaped valley the moraine may also include rocks which slipped down the valley walls; when the ice melted the moraine slumped downwards to the valley floor. Usually, it will have been added to by the post-glacial scree, which will also have accumulated at the foot of the valley wall.

laterite A reddish hard-baked soil which is found in tropical parts of the world. It results when hot wet conditions wash away most of the nutrient content of the soil, leaving behind hydrated oxides of iron and aluminium. The top is baked hard by the sun, and this factor, together with the leached nature of the soil, means that it is poor for agriculture. However, some laterites contain sufficient iron to be of commercial value. Laterites usually form when the soil has become exposed by forest clearance or by shifting cultivators who have not allowed the forest adequate time to recover.

latex A milky white fluid which is the source of rubber. It is obtained from various trees, especially *Hevea brasiliensis*, by tapping; that is, by making small cuts in the bark of the tree, through which the latex oozes. It is collected in a small cup which is attached to the tree just below the cut.

latitude The distance north or south of the equator, measured at an angle from the earth's centre. All lines of latitude are parallel to the equator which is the zero line of latitude. Each degree of latitude is approximately 69 mi. The tropics of Cancer and Capricorn are 23½° away from the equator, and the Arctic and Antarctic circles are at 66½°, which is 23½° away from the poles.

Laurasia The old continent which consisted of North America, Europe and Asia. Continental drift and plate tectonics caused the continent to split up into the separate continents which exist today. As the split occurred there was volcanic activity, for example in northern Ireland and western Scotland, similar to the present activity along the mid-Atlantic or Laurentia ridge in Iceland and elsewhere. See also **continental drift**.

lava Molten rock or magma which has reached the earth's surface as a result of volcanic activity. Lava varies in chemical content. If it is acidic it will be viscous and slow flowing, but if it is basic it will be more fluid.

Basic lavas form lava flows and often erupt out of fissures, as well as volcanic craters. Basic lavas are commonest at points where plate margins are moving apart, but acidic lavas occur at colliding plate margins and are often associated with very explosive volcanoes. Once they reach the surface, lavas cool and solidify. If the lava is very gaseous, a rough, jagged and clinkery surface will result, which is described as scoriaceous. See also **aa, pahoehoe**.

lava flow A movement of lava from a volcano or fissure. The movement may be fast if the lava is basic, but slow if it is acidic. A small lava flow can come from a volcanic crater, but sheets of lava may emerge from fissures.

lava plateau A large elevated area which has been built up by a series of volcanic eruptions, generally from fissures. A lava plateau covers several thousand sq km in the Deccan regions of India, and another large lava plateau is the Snake Colombia area of the northwest United States. A smaller example can be seen in the Antrim Mountains of northeast Ireland.

leaching The process by which chemicals and nutrients are removed from a soil. Rainwater, especially in warm climatic regions, will dissolve anything soluble and wash it down from the A horizon. Once removed, these solubles can be replaced only very slowly. Leached soils become coarse and are infertile. Large areas of the tropical lands have become seriously leached because of the removal of the forest cover.

least cost location An industrial site selected for ease of access to the necessary raw materials. This will save transport costs, and has been especially important in the iron and steel industry. Steel used to be made on coalfields to avoid transporting coal very far. Nowadays, because of changing technology, iron is the bulkier commodity. Most of the iron used in British steelworks has to be imported, and so many works are still near the ports that import the iron ore, for example Port Talbot and Redcar. Many steelworks in foreign countries are also situated along the coast, such as those in Baltimore and Osaka.

legume Any plant of the family *leguminosae*, such as peas, beans, clover and alfalfa, which extract nitrogen from the air and transfer it to their roots. When the plants die, the nitrogen is released in to the soil.

Legumes are sometimes described as nitrogenous; they are very beneficial to the soil, and are excellent crops in a rotation system.

leisure The use of free time is an increasingly important industry, and leisure has been a growth area of employment. Many leisure activities are located in towns; for example, sports centres, playing fields, entertainments, restaurants etc. Others are located in rural areas, where they sometimes come into conflict with farming. Walking may create problems of erosion on footpaths, as well as damage to fences and walls. Many areas are now being set aside specially for leisure activities; for example, country parks, theme parks, water sports, etc. Other important areas include national parks and heritage coastlines, which provide space and opportunities for leisure, whilst remaining agricultural land.

less developed countries The poorer countries of the world, where agriculture remains the major activity and employer of labour. The beginnings of industrial development can be seen in most countries, but it is often only in the capital city and a small number of towns. The less developed countries are sometimes referred to collectively as the South, and include much of Latin America, Africa and Asia.

leveche A hot dry southerly wind which blows into southern Spain. It is similar to the sirocco and khamsin, and blows in advance of a depression passing along the Mediterranean.

levée A natural embankment formed alongside a river by the deposition of silt when the river is in flood. Silt is deposited all over the flood plain, but most is deposited near to the river banks, and so a slightly higher area is created alongside the river. Levées can help to prevent flooding, and are sometimes built up and strengthened artificially. Some levées grow so large, that floodwater cannot return to the main channel. When this happens the river may change course, as often occurred on the river Hwang in northern China. Levées alongside rivers in Britain are only a few centimetres in height, but on some larger rivers, such as the Mississippi, they reach heights of 15 m.

ley Arable land temporarily sown with grass. Leys provide rich pasture for sheep and cattle.

liana Also **liane** A climbing plant which grows on trees in tropical forests. The stem of the liana may grow up to the top of the host tree, eventually killing it off. Lianas have been used as ropes by people living in the forests.

life expectancy The number of years a person may be expected to live based on statistical probability. Improvements in diet, hygiene and medicine have helped to increase life expectancy in most countries, although the more advanced countries are still much more favoured than the less developed countries. In Britain the life expectancy for men is about 72, and 77 for women, where as in several African countries life expectancy is only 45, although it is always slightly higher for women than for men. The highest mortality rate is before the age of one. This means that the life expectancy of anyone over one year of age will be higher than the national average. Life expectancy becomes greater with age.

light industry The type of industry which produces light, often small goods, that can be transported easily by road. Light industries are often located on industrial estates, or alongside main roads, such as by-passes and ring roads. Light industry is generally quite clean and non-pollutant, and can employ both male and female labour. Light industries often produce consumer goods, such as electrical goods, clothing and foodstuffs.

lightning A flash of light caused by a strong discharge of electricity in the atmosphere. The electricity is formed by tiny charges given off by large raindrops in a tall cumulonimbus cloud. Lightning occurs in thunderstorms, and it is the flash of lightning passing through the air which causes the thunder. The speed of light is much faster than the speed of sound, and so the lightning appears to precede the thunder. Lightning is visible on its way up from earth after it has struck. So, if you see lightning, it has already struck, and missed you. Cloud to ground lightning looks forked, but cloud to cloud lightning often creates a sheet of light.

lignite see **brown coal**

limestone* A sedimentary rock formed on the bed of a warm sea by an accumulation of dead sea creatures. At least 50% of a limestone consists of calcium carbonate. There are many different types including carboniferous limestone, chalk, dolomitic limestone, oolitic limestone, shelly limestone, Wenlock limestone. They all contain fossils or fragments of fossils. Because of the calcium carbonate content, limestones are soluble and permeable and generally give rise to dry landscapes. Chalk is pure limestone and forms rounded hills and vertical cliffs, if exposed along a coast. Chalk hills are often called downs or wolds. Carboniferous limestone is hard and forms high ground, such as the Pennines and Mendips. Oolitic limestone forms the Cotswolds and is often used as a building stone. Oolitic hills are generally more gentle than those of carboniferous limestone, but have a very pronounced west-facing scarp. Carboniferous limestone, like chalk and oolitic, has numerous dry valleys, but also contains caves with stalactites and stalagmites.

limestone pavement An outcrop of carboniferous limestone in which horizontal rocks have been weathered into clints and grykes. These look vaguely similar to the slabs of a pavement, though the cracks between the rocks are wider and deeper. Clints are rectangular in shape and may be up to two metres long, while grykes may be a metre or more in depth.

line graph A simple graph in which the values recorded are joined together by a continuous line, as for example, a temperature graph. It is a very clear method of displaying information.

literacy rate The percentage of people in a given population who have been taught how to read and write. The literacy rate ranges from nearly 100% in most European countries to less than 50% in several African countries.

lithosphere The solid layer which forms the surface of the earth. It rests on top of the mantle, and contains the sima and the sial. Between the lithosphere and the mantle is the moho. Also called **crust**. See also **mantle, Mohorovičić discontinuity, sial, sima**.

llanos (*Spanish*) Grasslands, especially the savanna-type grasslands of the Orinoco basin in Venezuela.

load The sediment carried by a river, ice, or the sea. The load will vary depending on the amount of available material, as well as the carrying power of the transporting agent. Large glaciers and icesheets can carry a greater load than a thin glacier, and will dump greater amounts of till when the ice melts. A river's load is greatest during times of flood, and

limestone

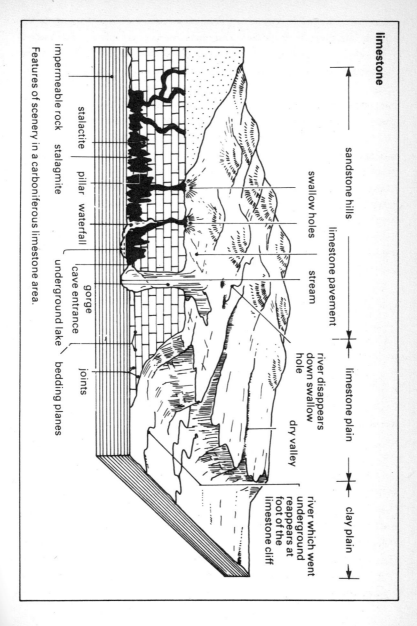

sandstone hills

limestone pavement

swallow holes

stream

river disappears down swallow hole

dry valley

river which went underground reappears at foot of the limestone cliff

limestone plain

clay plain

impermeable rock

stalactite

pillar waterfall

stalagmite

gorge

cave entrance

underground lake

joints

bedding planes

Features of scenery in a carboniferous limestone area.

139

quite small streams, which normally carry sand and mud, can move large boulders on occasions. The load may be in solution, suspension, or rolled along the bed. The maximum load will be determined by the volume and velocity of the river, and if the full load is being transported any additional material will result in some deposition. In times of flood, some of the load will be deposited across the flood plain.

loam A type of soil which is between a clay and sand in characteristics. It has the good features of both clay and sand, and is often rich in humus. It is normally a rich and productive soil.

local climate The weather conditions experienced in a small area, such as a valley, or a city. Local climate can also include studies of very small areas, but strictly this should be called micrometeorology. Micro studies consider variations within an area of 1 sq km or less, or changes below the height of the Stevenson screen.

location quotient A method for measuring the concentrations of a particular industry. Data for different regions has to be compared with information for the entire country. By use of the location quotient formula, it is possible to make a quantitative comparison between the different regions.

location theory A theory or idea which tries to generalize and simplify the location of various patterns of land use. Von Thunen's model for agriculture is one such example, since it locates, in theoretical terms, the most intensive types of farming next to the farm. The least intensive types of farming are located at the greatest distances from the farm, and the economic reasons for this can be explained quite simply. The Weber industrial model is another example.

lode A vein or seam of minerals. Often several different minerals occur along the same lode. Lodes were probably formed as a result of liquids or gases, heated by volcanic activity, forcing their way through existing rocks, and then cooling and solidifying. There are several lodes in the Peak District, where they are called **rakes**, and they contain lead, barytes, fluorspar, and other minerals. Some gold deposits in Australia and Alaska were in lodes.

loess Wind-blown soil which consists of very fine particles of dust. It is generally very fertile, like a loam, and is often very porous. Much loess is formed by winds blowing dust from deserts. The largest expanse is in northern China in the valley of the river Hwang, where dust from the Gobi desert has accumulated. As it is very soft, erosion can be quite rapid and the Hwang is sometimes called the Yellow River because of all the loess it transports. Erosion wears many small gorges in the loess, and local people sometimes live in caves dug into the walls of the valleys. There are several areas of loess in northern France, where it is called **limon**. Winds blowing from the icecap which lay further north during the last glacial phase probably picked up fine particles that had been deposited by streams of melt-water. There are small deposits of loess in southern England, which are called **brickearth**.

longitude The angular distance of a point on the earth's surface measured east or west from a prime meridian. Since 1884, the prime meridian of longitude has been located at Greenwich in London. Longitude is measured in degrees east or west of Greenwich, which is 0°. All lines of longitude are great circles, and they extend from the north pole to the south pole, intersecting the equator. As they all merge at each pole, they are obviously not parallel. A precise location of a place can be given by quoting its longitude and latitude, in the same way that the grid system is used on an ordnance survey map. Time zones are related to the lines of longitude, and 15% of longitude represents a time change of one hour.

longitudinal coast see **concordant coast**

long profile A longitudinal section of a geographical feature. A long profile is especially useful to show the shape of a river valley, from its source to the sea; it is a smooth and regular curve, except where broken by waterfalls, in contrast to the irregular long profile of a glacial valley. Also called **thalweg**.

longshore drift The movement of sand, mud and shingle along a coastline. The direction of drift can change, as it is caused by waves, which in turn are caused by wind. Most coastlines have a prevailing direction for longshore drift. Along the south coast of England it is normally from the west, and along the west coast of Wales it is normally from the south. Depositional material is moved up shore by the swash, which is at right angles to the crest of the waves. The material rolls back down the shore with the backwash, which is at right angles to the coastline. In this way particles are moved along the coast by successive wave movements. In order to reduce longshore drift, many towns have now constructed groynes at right angles to the shore. Sand piles up on one side of the groyne, but there will be little or no deposition on the leeward side of the groyne.

lough In Ireland, a loch, or lake.

low A depression or region of atmospheric pressure. In a low the atmospheric pressure is lower than in the surrounding areas, and it will be a region of ascending air and probably cloud.

low order goods see **convenience goods**.

M

magma Molten rock beneath the surface of the earth which forms igneous rocks when it solidifies. Below the surface, cooling is slow and large crystals form as the rock solidifies. Large masses of magma are in batholiths, and thin seams of magma create sills and dykes. If magma reaches the surface it flows out as lava. The different igneous rocks formed by magma vary according to chemical content, as well as depth of formation. See also **extrusive rocks**.

magnetism The earth can be compared to a huge bipolar magnet. Compasses point towards the North Pole or South Pole, depending on which hemisphere you are in. The magnetic poles are not quite located at the true poles, and their location changes slightly each year.

maize A widely grown cereal crop, *Zea mays*, bearing large spikes of cobs of grain. Maize is a staple food in many parts of Africa and in some parts of South America. It is grown in vast quantities in the United States, notably in the area of the mid-west known as the **corn belt**, where it is used primarily as a feed for cattle and pigs. Maize requires a wet growing season, with 500 to 750 mm of rain and needs summer temperatures of 20°C or more in order to mature and ripen.

malaria A widespread disease especially in tropical latitudes. It is transmitted by the anopheles mosquito, which breeds in stagnant water. It is proving difficult and expensive to eliminate malaria, but there are now medicines which can reduce the risk of catching the disease. Malaria used to occur in many more parts of the world, including southern Europe, and in recent years a more resilient type of mosquito has been extending the range of malaria across northern Africa.

mallee A type of scrub consisting mainly of low-growing eucalyptuses and occurring in Australia, especially in areas with a semi-arid or dry Mediterranean climate.

malnutrition Inadequate nutrition or nourishment. Malnutrition is not the same as undernourishment which is simply an insufficiency of food. Malnutrition is generally associated with a lack of vitamins, some essential minerals, and always a shortage of protein. It leads to poor health and a lack of energy, which often means that less money can be earned or less food grown. Several ailments result from malnutrition, including beriberi and kwashiorkor. Children are most seriously affected by malnutrition, and there are high infant mortality rates in several parts of the world, but especially in Africa, because of this problem.

Malthus, Thomas Robert (1766–1834) English political economist, who studied the growth of population. In 1798 he wrote down his conclusions; he believed that population was beginning to increase at a geometric rate and would double in 25 years. At the same time, agricultural productivity in the best farming areas could only increase food supplies at an arithmetic rate. This, he forecast, would lead to overpopulation and an insufficient supply of food. Malthus suggested that positive steps should be taken to overcome the problem. He advocated later marriages and observed that famine, disease and war would all check growth. His gloomy forecast was proved incorrect in the short term because of changes which he had not foreseen, but the problems he described have been relevant in the Third World during the 1970s and 1980s.

mangrove Any tree or shrub of the genus *Rhizophora*, characterized by the ability to survive in salt water. Mangroves grow on the shore. They produce masses of aerial roots, some of which anchor themselves in the mud on the seabed and send up new plants, thus extending the mangrove colony in a seaward direction. As the trees become denser, sediment is trapped round the roots and a swamp is created. Gradually the mangrove swamp becomes drier and firmer, as more muddy sediment accumulates. The drier parts will become inhabited by other plant species, while the mangroves continue to extend into new areas. Mangroves are found in tropical areas; for example, Malaysia, Indonesia.

man-made fibre see **ynthetic fibre**

mantle 1 A structural zone of the earth, situated between the lithosphere and the core. It is separated from the lithosphere by the Moho. The mantle is at an average depth of 30 km beneath the land, but only

10 km beneath the oceans. The rocks of the mantle are ultrabasic.
2 The surface accumulation of soil and weathered rock fragments.

map projections A system for representing on a flat surface the curved surface of the earth using a grid which corresponds to the lines of latitude and longitude. As the globe is round, and maps are flat, there is inevitably a certain amount of distortion when maps are drawn. Some projections give true shape or equal area; others may be based on the equator and will become less accurate towards the poles. One very popular projection is Mercator's which gives true shape as well as true directions.

maquis A type of scrub vegetation found in Mediterranean regions of France. It consists of low evergreen bushes and small trees, including myrtle, laurel, olive and arbutus, which have the capacity to survive the hot dry summers by means of water conserving devices, such as narrow or waxy leaves. They also have long roots to reach underground water supplies. Many of the plants are aromatic. During World War II, the French resistance movement adopted the name maquis, because it was always possible to hide in this fairly dense evergreen scrub vegetation. It is known as macchia in Italy, and is similar to the chaparral in California and mallee in Australia.

marble Limestone which has been metamorphosed by the effects of heat or pressure. The decorative and patterned effects in marble may be the result of the fossil content of the original limestone. Most marbles are light or whitish in colour, but the presence of small quantities of different minerals can produce a variety of other colours. Marble can be polished, which gives it much of its commercial value. Although there are limestones which can be cut and polished, they are generally not as decorative as marbles.

marginal land An area of land which is not very productive and needs to be farmed on an extensive basis. If economic returns are not good, it may not be worth farming at all. Marginal land is often on hillsides, such as the Pennines. To make the land more productive money would have to be spent on fertilizers etc., and the cash returns may not be sufficient to make this worthwhile. There are also marginal areas of semi-arid grasslands close to the wetter and more productive grasslands such as the prairies. Government grants are sometimes available to farmers in marginal areas as an encouragement to keep them in the area.

maritime air mass A large mass of air which has come from an area of sea. A tropical air mass will have come from the Atlantic before flowing over Britain, whereas a polar air mass will have come from the Arctic Ocean. The weather associated with the two air masses will be very different.

maritime climate A climate which is strongly influenced by proximity to the sea. Britain is an outstanding example, and has very mild weather for its latitude. The oceans warm up much more slowly than the land, because the heat of the sun is spread out through a great depth of water and ocean currents allow the heat to move vertically as well as horizontally. However, they also retain heat for much longer than the land. For this reason, climatic conditions near the ocean tend to be much warmer in winter and slightly cooler in summer than in inland areas in the same latitudes. This effect can be seen in Britain, though there are differences between the effects of the North Atlantic and the North Sea, the North Sea becoming much the cooler in winter, because it has little warmth from the North Atlantic Drift. In addition to the effects on temperature, the sea also influences precipitation, which is greater than in inland locations. The maritime climate of Britain is most marked in the Scilly Isles where the average temperature in winter is about 7°C – about the same as the Mediterranean coast of France. In summer the temperature averages 16°C, and the total annual rainfall is about 800 m. Further east in England the maritime effects are slightly reduced, and in Surrey the comparable figures are 4°C, 17°C and 625 mm. Moving further east into Europe, along similar latitudes, the climate gradually changes to continental; in Germany the corresponding figures would be about 0°C, 19°C and 550 mm. Outside Britain, maritime climates are found in Norway, northwest France, British Columbia in Canada, Washington and Oregon in the United States, southern Chile, Tasmania and South Island, New Zealand.

market area The catchment area for a central place. It is the area from which customers will travel in order to obtain goods and services. The size of the market area is determined by the size of the central place. Theoretically, market areas should be hexagonal in shape, but the real world is not quite as simple as the Christaller model.

market gardening see **horticulture**

marl A type of clay which contains large quantities of calcium carbonate and generally forms a productive alkaline soil.

marsh A soft wet area which suffers from poor drainage and frequent waterlogging. Many marshy areas are completely flooded for part of the year, becoming a little drier periodically. They are generally associated with areas of impermeable rock, and are often flat and low-lying areas of land, where the water-table is near the surface. Some marshy areas are also found in hill regions where rainfall is very heavy and peat bogs form; for example, the Pennines, Dartmoor. It is generally possible to drain marshes, by digging deep ditches and lowering the water-table, and this has been carried out in parts of the Fens and the Camargue in southern France. Draining changes the ecosystem completely and wildlife habitats can be destroyed. Conservationists normally wish to leave marshlands in their present state, whilst farmers often wish to drain them in order to make them productive.

mass movement The downhill movement of soil and rock due to the effects of gravity. Water can act as a lubricant, increasing movement after heavy rainfall or a period of snow melt. Freeze-thaw activity helps to provide loose material, which accumulates on sloping ground and then gradually slides downhill. There are various types of mass movement. Slow movements are called soil creep or rock creep. Faster movements are called earth flow or earth slide, and rock fall or rock slide, and these may be more dramatic and visible to the naked eye. Also called **mass wasting**. See also **soil creep, solifluxion**.

mass production The large-scale production of manufactured goods, often using an automated factory and a conveyor belt system. Machinery creating a large number of identical products can usually produce goods more cheaply than in small factories.

maximum thermometer A thermometer which records the highest temperature of the day. Usually found together with a minimum thermometer, it contains a small metal rod above the column of mercury. As the temperature rises the mercury moves up the tube, pushing up the metal rod. When the temperature starts to fall, the metal rod is left at the highest point which has been reached. It is then possible to read off the highest temperature later in the day or early the following morning, which is the usual time for taking temperature readings. By means of a magnet, it is possible to move the metal rod along to the mercury, in order to reset it. See also **minimum thermometer**.

meander* A bend or curve in the course of a river. A river will wind round any obstacle, such as hard rock, or even a pebble. Once a meander has been created it will continue, becoming accentuated by the erosive action of the river. On the outside of a bend there will be lateral corrasion, which will gradually work out sideways. Meanders gradually move downstream too. On the inside of the bend there is likely to be some deposition, which will build up a flat flood plain. Meanders tend to be quite small in the more mountainous upper course of a river, but they become larger further downstream. Where spurs still remain, there will be interlocking spurs, round which the river will meander. Down in the lower course, the meanders may spread over very large distances, and some on the Mississippi are up to 60 km in length. The erosion on the outside of the bend, especially in the upper and middle reaches will form a steep bluff and will gradually widen the valley. On the inside of the bend will be the slip-off slope, which is an accumulation of sediment. Meanders may eventually give rise to oxbow lakes. The term is derived from the name of a winding river (now Mendere) in Turkey.

mean sea-level The average level of the sea. Precise measurements are taken at Newlyn in Cornwall and other locations abroad in order to arrive at an official height. All other measurements are based on mean sea-level. The ordnance survey datum line used on ordnance survey maps is mean sea-level.

mechanical weathering see **weathering**

medial moraine see **moraine**

Mediterranean climate A type of climate associated with the countries surrounding the Mediterranean Sea, and also found in southern California, central Chile, near Cape Town in South Africa, and near Perth and Adelaide in Australia. Summers are hot and sunny, averaging 25°C or

meander

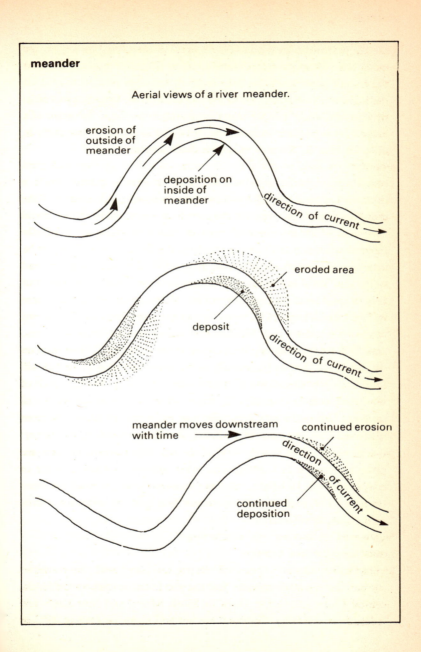

Aerial views of a river meander.

erosion of outside of meander

deposition on inside of meander

direction of current

eroded area

deposit

direction of current

meander moves downstream with time

continued erosion

direction of current

continued deposition

more; winters are mild, from 5° to 15°C. Summers are very dry, but winters can be quite wet, with up to 500 or 600 mm. Some of this rain comes as thunderstorms at the end of the summer.

megalopolis A very large city. The word is often used to refer to the area from Boston to Washington, including New York, Baltimore and Philadelphia. It can also be used for the San Francisco–Los Angeles–San Diego area, Honshu in Japan, and various locations in western Europe.

Mercalli scale A scale for measuring the intensity of earthquakes. It is numbered from I to XII; I indicates no damage and no real visible effects, although the seismograph records a slight tremor, while at XII on this scale there is total destruction and devastation, including the collapse of large buildings.

Mercator's projection A popular type of map projection, often used for showing the whole world. It is a rectangular map, based on a cylindrical projection. It was first used by Gerhard Mercator in 1569. All lines of latitude are the same length as the equator, whereas on a globe they become shorter towards the poles. At 60° north the line of latitude is only half the length of the equator on the globe. To balance this doubling of scale on the projection map, there is also a north–south doubling. This preserves accuracy of shape, which is one of Mercator's major advantages. However, there is considerable distortion in the high latitudes and Greenland looks much larger than Australia. The equal stretching of the land in all directions means that directions are accurately portrayed, and so the Mercator map is good for showing routes.

merry-go-round A system of transport in which trains or lorries perform the same journey repeatedly, sometimes on a circular route. It occurs commonly at power stations where there are direct links to a coalfield; a locomotive with a line of trucks goes through the loading hoppers at the coalmine and then moves to the power station, where it unloads onto the stock pile of fuel, before going back to the coalmine to load up again. The new coalmines at Selby supply Drax power station in this way.

mesa* A flat area of upland rather like a table, so called after the

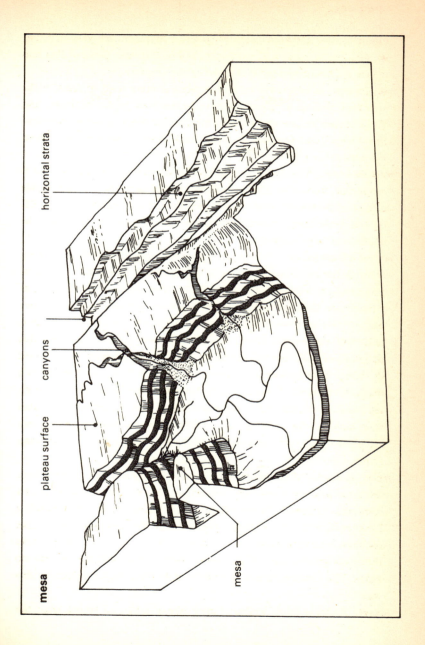

mesa

horizontal strata

canyons

plateau surface

mesa

Spanish *mesa* meaning table. Mesas are commonly found in arid or semi-arid areas, such as Arizona, Utah and New Mexico in the southwest United States. They often have horizontal strata of sedimentary rock, and their location in dry regions is also significant. The lack of rainfall means that the rivers are the major source of erosion. But they can cut deep valleys, which remain steep-sided, partly because of the horizontal structure, but also because there is little rainwash to wear away the sides and change the cross profile from U to V in shape.

Mesozoic The era of middle life, in the geological time scale. Mesozoic is between the Palaeozoic and Cainozoic eras. Palaeozoic is the time of primitive life, and Mesozoic is the time of reptiles. It lasted from 225 to 70 million years ago, and contains the Triassic, Jurassic and Cretaceous periods.

mestizo A person of mixed European and American Indian ancestry. When the first Europeans arrived in Latin America there were virtually no women amongst the early settlers; this led to intermarriage with the Indians. The early mixing of the races in Brazil and other countries has been a major reason for the lack of racial problems in South America.

metamorphic rocks Rocks which were originally igneous or sedimentary, but have been changed by the effects of heat or pressure, or both heat and pressure. Heat is the result of volcanic activity, and pressure is the result of earth movements. Metamorphic rocks are generally hard and resistant to erosion, and are likely to form high ground. In metamorphic rocks, minerals may recrystallize or be compressed, and the new rock may look completely different from the original sedimentary or igneous rock. Limestone becomes marble when metamorphosed; other changes include shale to slate, sandstone to quartzite, granite to gneiss. Examples of metamorphic rocks can be found in northwest Scotland, Charnwood Forest, the Malvern Hills and the Lizard Peninsula, and in large areas of Canada, Finland and Western Australia.

métayage see **sharecropping**

meteorology The study of the atmosphere. This includes pressure, temperature, clouds, wings, etc., and the description of weather and its explanation. Weather maps and weather forecasts are compiled by the Meteorological Office.

microclimatology The study of weather conditions in a small area and a short distance above the ground. The size of area studied may range from a few square metres to less than 1 sq km. However the study of urban climates may cover an area the size of London, but would still be considered a micro study. Microclimatology is sometimes defined as the weather below the height of the Stevenson screen. Variations between the weather at ground level and conditions a metre or so higher can be very important to plant growth. Exposure to wind, or shelter from it, can also be significant. See also **local climate**.

migration The movement of people (or animals) from one place to another. Migration may be permanent or temporary, and may be within one country, or from one country to another. People migrate in order to get away from something they do not like (**push factor**), or to go to somewhere which seems more attractive (**pull factor**). It will often be a combination of push and pull which makes people migrate. In Britain there has been migration to London and the southeast for many years, although there is now a slight reversal of this trend; in Italy the movement has been from south to north. Major migrations in the past included the movements from Britain and Western Europe to North America, Australia and New Zealand. More recently, many Jews have migrated to Israel, and workers from southern and eastern Europe have migrated to western Europe. Mass migrations of people from rural areas to the big cities have become a serious problem for many poorer countries, where shanty towns have sprung up around most large cities to house the increased populations.

millibar A unit of atmospheric pressure which is equal to one thousandth of a bar. It is used internationally on weather maps, on which isobars are drawn at 2 or 4 mb intervals. Average pressure in temperate latitudes is about 1012 mb; 1000 mb is equal to 29.53 in of mercury (750 mm).

millionaire city A city which contains more than one million inhabitants. The number of millionaire cities has risen rapidly since 1960 and there are now over 150. A large number of them are located in lowland regions, often near the coast, and many of them are capital cities.

millstone grit 1 A type of coarse sandstone which is best seen in the

Pennines. It forms high, rather bleak moorland, and in many places it is covered by thick peat bog. The poor acidic soil supports only poor vegetation, which can be used for sheep and grouse, and by walkers. 2 A geological age; the part of the Carboniferous period between the Carboniferous Limestone and the Coal Measures. It was the time when shallow seas covered parts of England and the sandy beach deposits were accumulating. In the Millstone Grit period, there are other rocks in addition to the coarse sandstone; fine sandstones, shales, thin limestones, as well as thin coal seams can also be seen. They were all caused by fluctuations of sea level at the time when the rocks were accumulating, about 320 million years ago. The rock was named millstone grit because it was used to make grindstones for mills; the shape of a grindstone has been taken as the symbol for the Peak District National Park.

mineral An inorganic substance with a particular chemical composition. Many minerals have a distinctive crystal shape; for example, cubic fluorspar or pyrites. Rocks are combinations of minerals.

minimum thermometer A thermometer which records the lowest temperature reached during the night. It does this by means of a metal rod which is placed inside the U-shaped tube with the mercury or alcohol. As the temperature falls during the night, the mercury or alcohol moves along the tube, pushing the metal rod ahead. As the temperature warms up again in the morning, the mercury moves back along the tube, leaving the metal rod at the coldest point, so that the minimum temperature can be read off. The metal rod can be moved along the tube with the help of a magnet. The minimum thermometer is usually found together with a maximum thermometer.

misfit A stream that is too small for the size of the valley in which it is situated. This may be the result of river capture, in which a river has been beheaded, and so is much smaller than it used to be. Many misfits can be seen in glacial U-shaped valleys, where the ice has made the valley deeper and wider than would normally be associated with the size of the river.

mist A cloudlike mass of minute droplets of water suspended in the air near the earth's surface. When air is cooled, it cannot hold as much water vapour. As the water vapour condenses, the drops become visible, forming mist. Mist is like thin fog, with a visibility of at least 1000 m. It is most likely to occur in calm, settled anticyclonic conditions, especially in autumn. It is generally very patchy, occurs most frequently near lakes, rivers and coastal areas, where there is more water available, and tends to accumulate in hollows. Once a wind starts to blow, turbulence will mix the air, and the mist will clear. In urban areas, where there may be pollution, the mist may become dark and dirty, to form smog. Smogs are now rare in London, because of the clean air legislation, but until the 1950s severe smogs occurred every winter.

mistral A cold northerly wind which blows down the Rhône Valley. It is caused by cold air from Europe blowing southwards into a low pressure area over the Mediterranean. The depressions occasionally pass along the Mediterranean from the Atlantic, especially during the colder half of the year. When the mistral blows down the Rhône, temperatures fall by several degrees, and the strong winds of up to 60 km per hr make conditions very unpleasant. Most farmhouses have rows of trees on their northern side as protection from the wind, and many fields of fruits and vegetables also shelter behind rows of cypress trees, which always extend from east to west and are very numerous in parts of the Rhône Valley. In the Camargue area of the Rhône delta, the landscape is very flat and windswept. When the mistral blows temperatures sometimes fall as low as freezing point, and many of the houses have doors and windows on the south-facing side only. A similar wind is experienced in other locations along the Mediterranean coast of southern Europe.

mixed farming Combined arable and pastoral farming. In the past, many farmers would rear animals and grow crops. The animals provided manure for the fields, which helped to maintain fertility. With crop rotation as well as animal manure, the structure of the soil could be maintained easily. In order to increase yields, many farmers turned to monoculture and had to spend large sums of money on fertilizer, which it is now realized can do much harm to the structure of the soil. Short-term gains in profit could be followed by long-term devastation or destruction of soil. Because of this, several farmers are now reverting to mixed farming, with crop rotation, and pasture and animals all forming part of the farming activity.

Mohorovičić discontinuity or moho The boundary between the lithosphere and the mantle, named after A. Mohorovičić who discovered it in 1909. The depth varies from place to place, but it is much deeper beneath the land masses than under the sea. The changing density of rocks at the moho causes earthquake waves to change speed as they pass it; from 21,000 to 27,000 ft per sec.

Mohs' scale A scale of mineral hardness, named after its inventor F. Mohs (1773–1839), a German–Austrian mineralogist. The scale is from one to ten.

1. talc – easily scratched by a fingernail,
2. gypsum,
3. calcite,
4. fluorite,
5. apatite,
6. felspar – the hardness of steel,
7. quartz – the hardest of the common minerals,
8. topaz – semi-precious stone,
9. corundum – used in carborundum stones for sharpening knives,
10. diamond.

monadnock An isolated hill, possibly the last remnant of a large range of hills which have been eroded for millions of years. A monadnock is a residual hill; it is probably a harder rock than the surrounding rocks, and so they have been eroded to a lower level. The name is derived from Mt Monadnock in New Hampshire, and there are many monadnocks in the southwest United States, where residual hills are often seen in Western films.

monocline A type of fold in which one limb (slope) is vertical or nearly vertical for a short distance, though the general trend of the limb of the fold may be quite gentle.

monoculture The growing of one crop. Monoculture has been a common practice for a long time in many places, such as plantations or vineyards, but it has become far more widespread during the last twenty or thirty years, especially in the production of cereals. This is as a result of the availability of machinery for harvesting large quantities very quickly. There are also many fertilizers which restore fertility to the soil, so that crop rotation is unnecessary. Monoculture has created problems, however, as dependence on one crop can be serious if the price suddenly falls because of a glut, or if new pests invade the crop. Moreover it is being realized that monoculture harms and possibly ruins the soil, in spite of the addition of artificial fertilizers. Monoculture in such areas as the prairies, the steppes and East Anglia is now declining slightly, and there are signs of reversal to crop rotation methods.

monsoon A seasonal wind, especially in south Asia. In summer, the monsoon season, the winds normally blow from sea to land and bring rain, but in winter there is a complete change of direction and the winds blow out from the land, giving dry weather. Some monsoon regions are very wet. Cherrapunji in India, for example, receives over 11,000 mm per annum, but others can be dry such as the Thar desert between India and Pakistan where the rainfall is less than 250 mm.

The major monsoon areas are in Asia, where the seasonal reversal of wind is greatest. This is because the largest continent, Asia, is adjacent to the largest ocean, the Pacific. In the smaller continents of South America, Africa, Australia and North America the monsoonal effects are less marked. These smaller continents do not have such wet summers or such dry winters, and are sometims referred to as eastern marginal rather than true monsoon. The rainfall brought by the monsoon conditions in Asia is cyclonic, but there is the addition of orographic (relief) rainfall in the mountains, which for this reason are often much wetter than the lowlands. In addition, the great heat during the day often gives rise to thunderstorms, and convection rainfall occurs.

moor A tract of bleak open land, generally on a plateau or mountain. The vegetation is likely to be heather or bracken, with some coarse grasses. Many moorland areas have extensive stretches of peat bog; for example, parts of the Pennines, Dartmoor. Acid soils often develop, especially in areas of heavy precipitation.

moraine An accumulation of boulders and rock fragments which have been deposited by ice. Some of the rocks may have been eroded by the ice; some may be the result of freeze-thaw activity; and others are weathered blocks that simply have been transported by the ice. Morainic debris is carried by the ice and dumped at the sides or at the end, or at the bottom of the ice, if it falls through the crevasses. Accumulations at the side of a glacier are called lateral moraines; at the end of the glacier they are called terminal moraines. When two glaciers merge, two lateral moraines unite in the middle of the enlarged glacier to form a **medial moraine**. Beneath the ice there will be gradual accumulations of ground moraine, but much of it is reduced to small fragments and powder by the effect of the ice grinding it on the bedrock. Ground moraine can become an effective abrasive tool, if scraped along the valley floor by the movement of the ice. Rock particles actually within the ice are described as englacial, but when the ice all melts, they fall to the floor and become part of the ground moraine.

morphology See **geomorphology, urban morphology**.

motorway A major roadway which has no sharp curves or junctions, and few turnings. There should not be any sudden obstacles on these routes, which are designed for high speed. There are speed limits on most motorways; for example 70 m.p.h. in the United Kingdom. Motorways first appeared in Britain in the late 1950s, but there have been Autobahnen in Germany since the 1930s. In many countries drivers have to pay a toll to use the motorways.

mulatto The offspring of a white person and a Negro.

multinational A large company which has interests in several countries. Many multinationals have become very influential in one particular product, and may control production and marketing in certain countries. This may be beneficial, but there are certainly situations in which a multinational may take a course of action which benefits the company but does not really help the local people or the country. It is a problem which affects Third World countries in particular. Tea and bananas are produced by multinational companies, which can influence developments in the growing areas. They are often accused of using cheap labour and exploiting the local people; on the other hand, they claim that they are helping development in the area by providing jobs, and also schools and hospitals on occasion.

multiplier effect   The way in which the creation of one or two industries may encourage the setting up of other industries in the same area. For example, a steelworks may provide raw materials for an engineering works, which may provide machines for a textiles factory, and so on. Once started, the effect will continue, because there will be a need for transport, schools and hospitals, which will provide more jobs and also create a good local market. More people will be encouraged to move in to the area, and possibly additional industries will be attracted. Soon administration will be necessary, providing yet more jobs.

multiple nuclei model   One of the classical models of urban morphology, which shows that large cities develop and grow round two or more separate central places, or nuclei. The nuclei may have separate settlements which have been swallowed up by urban growth. Harris and Ullman wrote about this model in 1945, and they believed it added important ideas to the models described by Burgess and Hoyt.

mushroom rock   see **pedestal rock**

muskeg   A boggy area which contains sphagnum moss, and other mosses and lichens. It is found in the tundra regions of Canada, and also in the northern parts of the coniferous forest belt.

Myrdal, Gunnar   A Swedish economist who in 1957 wrote about his theory of cumulative causation. He believed that the developments of industry, agriculture and economic growth would occur most readily in areas where there were already successful activities. Success would breed success and attract other forms of development. His ideas fit in with those of Friedman, who formulated the core-periphery idea, and in many countries attempts have been made to establish new core areas.

N

nappe see under **folding**

national park A specially designated area of land, where development is controlled and planned. In England and Wales the land is not really parkland and is not owned by the nation. The eleven national parks of England and Wales are mostly farmland, and are located in scenically attractive areas. There have been ten national parks for many years, and the Norfolk Broads was added in 1989. There are no national parks in Scotland, but there are forest parks, which have many similarities in planning controls and the provision of amenities. Other countries have national parks, which are much bigger than those of the United Kingdom; for example, Yellowstone National Park in the United States. National parks are usually left in a wild and natural state, and may contain many wild animals, high mountains, geysers, glaciers, etc. They generally cater for visitors, with hotels and camp sites, and provision for fishing and walking. Most national parks in the world aim to preserve or conserve the landscape, and to enable the public to enjoy the countryside.

natural arch An archway formed by marine erosion. On an exposed headland, a cave may be enlarged until it cuts through to the other side of the rock to form a tunnel or an archway; for example, Durdle Door in Dorset.

natural bridge A bridge of rock formed by erosion in a limestone area. When an underground stream erodes a tunnel and forms a large cavern, if the roof collapses a gorge will result. In locations where two caverns have collapsed quite close to one another, a natural arch or bridge can be left behind; for example, Natural Bridge in Virginia.

natural gas A combustible mixture of hydrocarbons in the form of ethane and methane found in trap rocks in the earth's crust, often together with oil. The gas will need treatment or purification, but can be used as a very clean source of energy. Britain's main sources of natural gas are in the North Sea; they have been utilized since about 1970. Natural gas is an important source of energy in many other countries, including the United States, Soviet Union and the Netherlands.

natural increase The increase in population caused by the difference between the birth rate and the death rate. When the birth rate is 40 per thousand and the death rate only 20 per thousand, the natural increase is very high. This was true of Britain about 200 years ago, and is now typical of many Third World countries. The overall position, however, depends on immigration and emigration as well as natural increase. See also **zero population growth**.

natural vegetation The kind of vegetation which would grow naturally without interference from man. There are probably very few parts of the world where truly natural vegetation remains, as most areas have been cleared, even if there has then been a secondary growth of natural vegetation. Parts of the tropical forests and some areas in the Arctic or on high mountains may still be as they were hundreds of years ago. However, vegetation does alter naturally, and Britain, for example, has changed from icecap to tundra, to coniferous forest, and then mixed woodland since the ice disappeared from the land about 11,000 years ago.

nautical mile A unit of distance used in navigation. Slightly longer than a normal mile, it represents one minute along the line of a great circle (60 minutes equal one degree of the earth). The distance is about 6080 ft (1853 m).

neap tide A tide in which there is minimal variation between high tide and low tide as a result of the lunar and solar tidal influences pulling against one another. The opposite is a **spring tide**. Spring tides are the result of lunar and solar tidal influences working together to create a high tidal range. Neap tides and spring tides happen every month, the spring tide occurring two weeks after the neap tide. Neaps are the result of the effects of the sun and the moon offsetting one another and balancing out their influences.

nearest neighbour analysis A technique for describing the distribution of features, such as churches, windmills, etc. The formula gives a precise quantitative answer, which will be between O and 2.1. The answer will show whether the distribution of the features is random or regular, etc., and the figure for one area can be compared with that of another.

network An interconnecting pattern of roads, railways or the like. It is often easier to draw communication networks as topological maps, which look neater and often clearer, for example, the London Underground map.

névé (*French*) see **firn**

newly industrializing countries Any of the less developed countries which have built up their industries in recent years. Outstanding examples are Taiwan and South Korea in the Far East. Brazil, Nigeria and many other countries have begun to use the wealth of their natural resources for industrial development. Because of low labour costs they can sometimes undercut the old established industrial regions.

New Red Sandstone The Permian and Triassic geological periods. See **Palaeozoic**.

new town Any of various towns in Britain which were planned and built as a whole. Most new town development has taken place since World War II. The Abercrombie plan for Greater London suggested the creation of a ring of new towns about 50 km from London. They were designed to attract people away from London and restrict its further expansion and, although this did not happen, the new towns grew into thriving communities, such as Bracknell, Crawley and Hemel Hempstead. In the 1960s further new towns were developed in some of the old coalmining and heavy industrial areas where there had been serious industrial decline and high rates of unemployment. There developed such places as Cwmbran in south Wales, Washington in County Durham, Cumbernauld in central Scotland, Skelmersdale near Liverpool and Telford, which includes Ironbridge and Coalbrookdale. They all have trading estates as industrial locations, and a variety of light and consumer goods industries. They provide overspill accommodation for the population of large cities, and in some cases, such as Telford, they become centres for regional development. Most recently there has been the rapid growth of Milton Keynes, as an outstanding example of a new town.

nick point see **knick point**

nimbostratus A low layer of dark clouds sometimes occurring at the warm front of a depression.

nimbus cumulonimbus. See **convection rainfall**

nivation 1 Mechanical weathering caused by freeze-thaw activity. Water lying in the cracks in rocks freezes and expands. As the temperature rises, the ice melts, only to freeze again the following night. The repeated freezing and thawing, expanding and contracting, causes pressure and stress on the rocks, which eventually crack along any lines of weakness. Nivation causes the formation of angular fragments of rock, which accumulate at the foot of rocky outcrops. Accumulations on slopes form scree. Examples can be seen on hills in Britain, as well as in the Alps and Arctic environments. 2 The freeze-thaw activity which takes place around the edges of a snow patch. It may erode particles of rock to form a slight hollow, in which snow may accumulate again the following year. Nivation hollows may eventually become corries if the freeze-thaw activity continues for many years.

node A point in a network where routes meet or intersect, such as a settlement or crossroad.

nomad A wanderer, a person with no fixed home. Most nomads nowadays are only semi-nomadic, as they have a home village in which they spend several months each year before going off on their wanderings. Some nomads wander in search of food for themselves, but most are pastoralists, who have to take their animals to different areas in order to find pasture. Nomadic pastoralism is a system of farming found in many different locations, including temperate grasslands, tropical grasslands, semi-deserts and tundra. In the tundra there are reindeer herders, such as the Lapps of northern Sweden and Finland; in central Asia there are cattle and sheep herders; and in the Sahara there are camel herders. Just to the south of the Sahara is the Sahel region. Here there are many pastoralists such as the Fulani. Most groups rear cattle though some keep goats. They wander round for several months each year seeking pasture for their cattle. There are now more people and more cattle than there used to be, and so much of the sparse pasture is being overgrazed. This has led to desertification, and many areas which used to contain some grazing are now quite useless. Some pastoralists try to find feed for their animals further south, where it is wetter and more vegetation grows. Many new water-holes have been dug, and irrigation schemes have been started in order to help these people. In most developments the nomads are expected to settle permanently, so that in some areas nomadism will gradually disappear.

non-renewable resource A resource which cannot be replenished. Fossil fuels, for example, cannot be replaced or reused once they have been burnt. The same is true of metals, although some scrap and waste metals can be collected and recycled. Metal ores are a finite resource which cannot be renewed.

North, the A region including the more advanced areas of the world – Europe, North America and Australasia – as defined by the Brandt Commission.

North Atlantic Drift see **Gulf Stream**

northing The east-west grid line on an ordnance survey map. It is quoted second, after the easting, when giving a grid reference.

nuclear power station An electricity generating station which uses nuclear fuel as the source of energy. There are several different types of nuclear power station, and although they are expensive to construct they are quite cheap to run. They do not give off pollution like the thermal power stations, but they do present serious problems about the disposal of radioactive waste. Most nuclear power stations are near the coast in Britain, as large quantities of water are required for cooling purposes, and many have been built in fairly isolated locations.

nucleated settlement A small settlement which has grown up around a nucleus, such as a village green or crossroad. Nucleated settlements are a cluster of dwellings; they are not spread out along roads in a linear fashion. Nucleated settlements often grow at a spring or water-hole, or on a higher patch of dry ground in a marshy area.

nuée ardente A cloud of hot gas which comes swiftly down the side of a volcano after an eruption. The cloud may also contain small dust particles and fragments of ash. It will kill anyone it passes over; in the Mt Pelée eruption of 1902, all the inhabitants of St Pierre were killed by the nuée ardente. There was also a nuée ardente at the eruption of Mt St Helens in 1980.

null hypothesis A statistical method used to determine the significance of the differences between two samples, or to frame a question which is likely to be proved wrong by field studies or research. A null hypothesis is a negative assumption, when the researcher really believes that a positive answer is likely. Research may show that a null hypothesis is false, although it will not necessarily produce the correct answer.

nunatak The eskimo word for a rocky outcrop projecting above the general level of an icesheet, as in Greenland and Antarctica. It may be the top of a mountain peak.

nutrient cycle The transfer of nutrients in an ecosystem from one stage to another. Leaves fall to the floor to be broken down by bacteria and provide nutrients for the soil, which then sustains plant growth. The growing plants take energy from the sun, and provide food for any animals which live in the ecosystem.

O

oasis A fertile location which has water in an arid landscape. Usually ground water is brought to the surface in a well, but an oasis may occur at the point where a river flowing from a wetter region crosses the desert on its way to the sea, such as the Nile or Indus. Most Saharan oases contain large numbers of date palm, and there may also be many fields of cereals, vegetables and fruit. There is often a settlement in the oasis, sometimes with up to 30,000 inhabitants.

oats A cereal plant, *Avena sativa*, which can be used as a human food, in breakfast cereals and porridge, and also an animal feed. The plant can survive in fairly wet environments, such as western Scotland, where rainfall totals may be over 1000 mm, though the grain will not ripen in such locations. Oats can tolerate low temperatures in winter, and requires summer temperatures of up to 15°C. Oats is widely grown in the uplands of Britain, and in Norway, Sweden, and other northern latitudes.

obsidian see **volcanic rock**

occluded front A combination of a warm and a cold front in a depression. It occurs when the cold front catches up with the warm front and undercuts it. The effects of the warm front and the cold front together are to cause heavier and more prolonged rain than either front would give alone. Cumulonimbus clouds will be found at the occluded front. An occluded front is sometimes described as an **occlusion**. A large number of depressions have occluded by the time they reach Britain.

ocean The oceans cover about 71% of the earth's surface. The Pacific is the largest, and it contains the deepest part of all the oceans. The others are the Atlantic, the Indian, the Arctic and the Southern Ocean, which surrounds Antarctica. The continental shelves are the shallow parts of the oceans, and occur nearest to land. At the end of a continental shelf is a continental slope leading down to the deep sea plain, which covers vast areas between 4000 and 6000 m deep. The ocean floor has many ridges and trenches, and in the trenches are the ocean deeps, about which still very little is known.

ocean currents A distinctive and persistent movement of surface water. Most currents are the result of winds blowing the surface water. In the Atlantic there are equatorial currents, which flow from east to west because of the effects of the trade winds. The North Equatorial Current is just north of the equator and it flows into the Gulf of Mexico. It turns to the right because of the shape of the land and the effects of the rotation of the earth, and then flows across the Atlantic as the North Atlantic Drift. The current goes to Britain and Norway, but a mass of water turns right to head southwards, past Portugal and along the African coast. At this point it is called the Canaries Current and is a cool current, not because the water is really cold, but because it is cool for its latitude. For the same reason, all currents which flow towards the equator are cool currents. The Canaries Current rejoins the circulation to become part of the North Equatorial Current again. The other major current which joins this clockwise circulation is the cold Labrador Current; it heads southwards along the west coast of Greenland, bringing icy water from the Arctic. It also brings icebergs, as far south as Newfoundland on occasions. In the South Atlantic there is a similar though anticlockwise circulation, and the cool current off the coast of southwest Africa is the Benguela. In the Pacific there are circulations which resemble those of the Atlantic. The Indian Ocean is similar, but halfway through the year there is a reversal of flow, because the monsoon winds change direction. In summer, while the south-westerlies are blowing, the current flows from west to east, but in winter the winds change to north-easterly, and the ocean currents flow from east to west.

ocean-floor spreading A movement of the sea-bed associated with plate tectonics. Along the mid-Atlantic ridge, for example, two plates are moving apart, and as a result a split is developing through the middle of Iceland. As the plates move, fissures open up and basic lava pours out; for this reason Iceland is the most active volcanic area in the world at present. In the Mesozoic era, western Europe and North America were close together, but have been drifting apart for millions of years because of the ocean-floor movements. Ocean-floor spreading is also taking place near Aden and Somalia and between the Nazca and Pacific plates off the west coast of South America. Also called **sea-floor spreading**.

oil terminal A port where oil tankers carrying oil from producing regions unload in the importing country. Oil terminals have facilities for storing or refining the oil. Milford Haven is a major oil terminal in south Wales, and London, Southampton, Merseyside and Teeside are also important. Rotterdam is the largest terminal in Europe.

okta (in meteorology) A measurement of cloud cover equal to one-eighth: eight oktas means the sky is completely covered and four oktas that it is half covered.

O layer see under **A horizon**

onion weathering see **exfoliation**

oolite A type of limestone consisting of calcareous shells which accumu-lated on the sea bed. Oolitic limestone was formed in the Jurassic period. Also called **oolith**.

open-cast mining Mining of rocks or minerals which are on or near the surface. There is often a layer of rock, called **overburden**, on top of the mineral, which is removed and placed on one side, so that it can be replaced at the end of mining operations. Digging and excavating equipment are then used to remove the required mineral. Open-cast is generally much cheaper than shaft mining or adit. It is frequently used for coalmining in Britain, and to extract lignite in Victoria, Australia and iron ore at Kiruna in Sweden.

optimum population The best population density for a particular region. It means there will be jobs for most people, and ample income and food supply. Some places have too many people for the available resources; for example, many parts of India, especially towns such as Calcutta, are overpopulated. Other regions may have fewer people than could be supported at a reasonable standard of living; for example, much of Australia. The optimum population of an area can change, if there are changes in sources of materials and income. For example, if all the oil, natural gas, and coal in Britain were exhausted, the optimum popula-tion would become far lower than it is at present.

ordnance datum The height from which measurements are taken; mean sea-level.

Ordnance Survey The organization which produces the detailed maps of all parts of the United Kingdom. The Ordnance Survey was created by the War Office and had a military basis, though it is now a civilian organization, with headquarters in Southampton.

ore A type of mineral or rock which contains a metal or non-metallic substance of some commercial value. Iron ore is likely to be a type of sandstone with a percentage of iron; if it is 25%, then of every 100 tn of rock dug up, 25 tn will be iron. Good quality iron ore is 60% to 70% pure, whereas a good quality copper ore may be no more than 1% pure.

organic Derived from the breakdown of vegetation or animal matter. The organic part of soil is the remnants of living matter, whereas the inorganic content is the particles derived from the rocks.

organic fertilizer Any fertilizer derived from organic material, such as manure, bonemeal, seaweed, or fishmeal. It is thought to be better for the structure of the soil than chemical or inorganic fertilizers.

orogenesis The process of mountain formation. As a result of movement of the earth's plates, folding and faulting are likely to occur, as the crustal materials are compressed and forced up. The major periods of mountain formation have been the Caledonian, Hercynian and Alpine. The Caledonian were formed at the end of the Silurian period, and good examples can be seen in Scotland and Norway. The Hercynian folding occurred at the end of the Carboniferous period, and formed the Harz mountains, the Massif Central in France and the Meseta in Spain. The Caledonian and Hercynian are referred to as old fold mountains. The Alpine folding took place during the Tertiary era; mountains formed at this time are referred to as young fold mountains. Also called **orogeny**.

orographic rainfall Rainfall which occurs as a result of winds rising to pass over high ground. As the winds rise the air is cooled causing condensation, and clouds form. If the air continues to rise rain will fall. In temperate latitudes the total rainfall will increase with height, up to about 2000 m. Above that height, the air becomes so cold that it cannot hold as much water vapour, and so rainfall quantities begin to decrease.

There is nowhere in Britain which exceeds this height, and so it is true to say that in Britain rainfall increases with height. The winds which cool as they ascend mountains warm up as they descend on the leeward side, and are able to hold more moisture. So, the rainfall totals are much lower on the leeward side of the hills, and it is referred to as a *rain shadow*. All mountains are likely to have an area of rain shadow. In England it is on the east side, but in Australia it is on the west, So that while the area of Queensland to the east of the Great Dividing Range receives 800 to 2500 mm annual rainfall, the west side totals 250 to 500 mm per annum. Also called **relief rainfall**.

orthoclase   see felspar

outcrop   A part of a layer of rock or rock formation which is exposed at the surface of the ground.

outlier   A mass of newer rocks surrounded by older rocks, often because it has been detached from a larger formation by erosion. There are several outliers to the west of the main Cotswold scarp.

out-of-town location   A site in a rural area near to a city, in which new development can take place. Several new industries have been created in out-of-town locations, where there is more space for building, land may be cheaper and there may be less traffic congestion. Many new hypermarkets and supermarkets have been built in out-of-town locations, where there is plenty of space for parking.

outport   A secondary port which is downstream nearer the sea than the main port. Sometimes the main port has declined because modern ships are unable to get as far upstream as the smaller boats of the past. Tilbury is really an outport for London and Bremerhaven for Bremen.

output   The end product of a process or system; for example, the crops or animals produced on a farm. See also **input**.

outwash plain   A depositional area built up by the sands and gravel brought down by streams flowing from a glacier or an icesheet. The materials have literally been washed out by fluvio-glacial activity. They are often sorted, and so the largest particles are deposited first, the materials becoming finer and finer moving downstream. Broad plains, called **sandur** in Iceland, are sometimes built up by several streams flowing from an icecap. There are outwash deposits in Norfolk and on the corn belt of the United States. They are often sandy and infertile.

overburden see **opencast mining**

overflow channel A valley formed by water overflowing from a lake. There are several examples in the Pennines, where meltwater from the icecaps formed a lake, which then cut out an overflow channel in the lowest of the surrounding hilltops. In the North Yorkshire Moors there is a well-known overflow channel called Newtondale. Meltwater formed a lake at the northern edge of the moors, but found its easiest outlet by cutting a route to the south. At Ironbridge a deep gorge was formed by overflow water from a lake to the north. The lake was created as a result of the upper parts of the Severn becoming dammed by ice.

overfold A fold in which the rock strata have been pushed right over because of intense compression.

overgrazing The excessive use of an area of land for pasture. If too many animals are allowed to graze an area, they eventually remove all the grass and leave the soil exposed. Once this happens the soil is likely to be eroded, and no more grass will grow. Overgrazing can destroy the land and turn it into desert. This has been happening in the Sahel, with the result that the Sahara is spreading southwards. Cattle and sheep can cause overgrazing, but goats are even more notorious for removing all the vegetation and leaving the soil bare. To bring about a reduction in overgrazing is difficult, because animals are the only source of wealth, food and income for many people. More feed could be grown for the animals if irrigation schemes were set up, and this would enable some of the pastures to be rested and restored. Rotation of pastures would be vital in areas where overgrazing has taken place.

overpopulation The condition of having too great a population for the available resources. If a country or a region is unable to maintain a reasonable standard of living for its inhabitants, it can be described as overpopulated. It is not necessarily related to the density; some parts of the United States, for example, have high densities of population, but since everyone is relatively wealthy and well fed, the population is not too high. On the other hand, parts of the Sahel are quite sparsely populated, but the land is too poor to support the population. The situation could be reversed if, for example, better farming techniques were introduced. Equally, there could be a reversal in wealthy countries. If all the minerals ran out and industry declined, then they could reach a state of overpopulation. See also **optimum population**.

overspill A proportion of the population of a large town or city which moves out to find more space or better living conditions in a new town a few miles away. Sometimes a new town is created to cater for the overspill, or an existing town may be enlarged, such as Redditch, near Birmingham.

oxbow A small lake which was originally a meander of a river. As sediment is deposited, the meander becomes cut off from the river to create a lake. Once formed, the oxbow will gradually shrink as sediment fills it in; vegetation will grow on the new muddy area, and the land can be reclaimed. Also called **cut-off, horseshoe lake**.

ozone A form of oxygen which is found in very small quantities in the atmosphere, mostly in the ozone layer, about 20 km above the earth's surface. The ozone layer absorbs much of the ultraviolet wave energy, but allows longer ultraviolet waves to pass. Gaps have been seen in the ozone layer in recent years, and this could cause serious changes in climatic conditions. The additional ultraviolet waves which could get through to earth might also constitute a health hazard.

P

Pacific coast see **concordant coast**

packet port A small port which carries passengers as well as mail. The cross channel ports, such as Dover, Folkestone, Newhaven, are sometimes called packet ports.

padi Also **paddy** 1 The rice plant, *Oryza sativa*. 2 A wet field in which rice is grown. Not all rice is grown in padi fields, since some varieties can be grown in quite dry areas. See also **rice**.

pahoehoe The Hawaiian name for **ropy lava**.

palaeomagnetism The study of old magnetic fields and the magnetism which can be found in ancient rocks. When igneous rocks cool and harden they retain a certain magnetization, resulting from the presence of particles of iron oxide, which orientate themselves parallel to the earth's magnetic field at that period. This enables geologists to work out the location of these rocks at the time of their formation and can be very helpful in studies of drifting continents and plate tectonics.

palaeontology The study of fossils. It is more relevant to geology than geography, but fossils are seen in studies of chalk and limestone rocks, and also in coal measures, where there are many plant fossils. The study of fossils, and the plants and animals which created them is an important way of finding out conditions on earth millions of years ago, since the requirements for growth of the different plants and animals are known. Fossils have been found on Mount Everest which reveal that the rocks were in a warmish sea about 40 or 50 million years ago.

Palaeozoic The geological era which lasted from about 600 million years ago until about 225 million years ago. It includes six periods: Cambrian, Ordovician, Silurian, Devonian, Carboniferous and Permian. During the Cambrian, Ordovician and Silurian periods much of Britain was covered by sea, and trilobites, graptolites and brachiopods were the commonest animals. Then the Caledonian orogenesis took place, and new mountains were formed at the beginning of the Devonian period. As these mountains were eroded, the sea spread over Britain again to form the Carboniferous limestone, in which corals and crinoids were common animals. At the end of the Carboniferous period the Armorican or Hercynian orogenesis took place, and Britain became land for the Permian period which, together with the Triassic period of the Mesozoic era, was known as the **New Red Sandstone**. The word Palaeozoic means ancient life, and the era contained fairly primitive animals only. It was followed by the Mesozoic era. Also called **primary era**.

pampas An area of temperate grassland near the Plate estuary, lying mostly in Argentina and also in Uruguay. The name is derived from the Spanish word for plain. With temperatures of 20° to 25° C in January and 10° to 15° in July and an annual rainfall of 500 to 1000 mm conditions are ideal for grass. Early settlers reared sheep and cattle on the rich pastures, and there were local cowboys, called gauchos, to look after the herds of animals. Nowadays, much of the grassland has been ploughed, and there are vast fields of maize, flax, wheat and alfalfa. Large cities such as Buenos Aires and Montevideo have grown up in this region.

Pangaea The name given to a vast land mass which millions of years ago was the only land on the earth. It consisted of the sial which gradually split to form the ancient continents of Gondwanaland and Laurasia. Pangaea began to break up about 200 million years ago; it was first written about in 1912 by Wegener in connection with his theory of continental drift. See **continental drift**.

pastoral farming The rearing of herbivorous animals. In addition to grazing on pasture, many animals have to be fed, especially in the winter months. There are a number of different types of pastoral farming, including sheep-farming in the Welsh hills, the Pennines, or the Murray Darling basin of Australia; dairy farming, as in Denmark and New Zealand, or in Devon and Cornwall; and the rearing of beef cattle on the plains of Texas. Some pastoral farmers in parts of Africa, such as the Masai or Fulani, are nomadic and wander around looking for grazing for their animals. In other regions there are pastoralists who rear reindeer, llama, yaks, or goats.

pasture An area of grassland suitable for grazing by animals, such as sheep, cattle, goats, llamas, yaks, alpacas, or deer. Most important and most numerous are sheep and cattle. Sheep can survive in areas of poor pastures, such as hilly or semi-arid areas. They do not produce good meat in such environments, but they yield fine quality wool. Sheep also thrive in richer pastures and wetter areas, where good meat as well as wool can be produced. Dairy cattle are found only in areas of rich pasture, as they need to be well fed at all times, but beef cattle can live in poorer areas. Pastures in temperate humid landscapes such as in Britain are generally very nutritious for animals. Other areas such as the pampas are also very suitable, but drier areas, such as the high plains of Kansas, Wyoming, and elsewhere in the western United States, can have very poor pasture. In tropical lands the grassland is often quite rich in summer, but very dry, brown and shrivelled in the winter months.

peasant farming Small-scale farming, in which a farmer grows crops and rears animals mainly for his own use, but probably with a small surplus for sale. It is intermediate between commercial and subsistence farming, but nearer the latter. Some peasants are very poor, as in parts of west and east Africa, but others may be leading quite a comfortable life; for example, in many parts of France. A peasant farmer usually has very close links with his land, and looks after it very carefully.

peat A soil composed of dead and decaying remains of vegetation. It is generally dark brown or black in colour, and is thought to be similar to the first stage in the formation of coal. It forms in areas of high rainfall or very poor drainage, and contains a high proportion of water. Peat is common in upland areas of the millstone grit Pennines, Dartmoor and many parts of Scotland. It also occurs in lowland areas, such as the Fens, Somerset Levels and much of Ireland. In very wet conditions, plants do not decay when they die, because the conditions are nearly anaerobic, and so the vegetation is not broken down by bacteria. Waterlogged and airless conditions create a very acidic and infertile landscape, which can become very boggy. Only specialized plants, notably sphaghnum moss, can survive in such areas. Large tracts of upland Britain are covered by blanket bog, which is a large sheet-like expanse. The low-lying peaty areas such as in the Fens can be drained more easily, and the black soils can be quite fertile and productive. This is because the plants which formed the lowland peat are less acidic and contain more nutrients.

pedalfer   A large group of soils which contain aluminium (al) and iron (fer). The zonal classification of the world's soils contains two groups, the pedalfers and the pedocals. Pedalfers are found in the more humid regions, such as many parts of the tropics, where **laterites** are a common type of pedalfer. In maritime regions such as Britain, brown forest soils are common, and in the cooler coniferous forests of Canada, Norway, Sweden, Finland and the Soviet Union there are widespread **podsols**, another common type of pedalfer. See also **pedocal**.

pedestal rock   A rock which is shaped like a mushroom as a result of corrasion by sand. In arid landscapes, windborne sand is moved along just above ground level. The sand erodes any rocks which are standing above the surface of the land, and as the corrasion is most active a metre or so above ground level the eroded rocks become mushroom- or pedestal-shaped. Pedestal rocks occur in most deserts; examples can also be seen near some coasts and on hills such as the millstone grit areas of the Pennines. Also called **mushroom rock**. See also **zeugen**.

pediment   An area of eroded fragments of rock close to uplands in a semi-arid or arid region. The accumulated fragments gradually build up a plain which slopes down gently from the mountains. The largest rocks and boulders are generally found nearest to the mountains, and the particles become progressively smaller with distance away from the high ground.

pediplain   A large expanse in a desert or semi-arid region which is fairly low-lying and flat. It is formed by the merging of several areas of pediment.

pedocal   Any of a large group of soils which contain a high proportion of calcium. Pedocals are found in dry regions, such as the drier parts of the steppes and prairies, and the semi-arid lands of the world. They are one of the two main groups of zonal soils, the other group being the **pedalfers**. Pedocals do not suffer much from leaching, because they only occur in areas where the annual precipitation is about 500 mm or less.

pedology The study of soil.

pelagic Occurring in or inhabiting the open sea. Certain fish are described as pelagic fish because they live out in the moderately deep parts of the ocean as opposed to the shallow areas of continental shelf. Pelagic deposits are fine muds which accumulate on the bed of the oceans. They consist mainly of the dead remains of tiny sea-dwelling plants and animals. Remains of this kind will also fall to the seabed in shallower waters, but are lost among the greater deposits of land sediments brought to the sea by rivers, flowing from adjacent land masses. The fine muds are called **oozes** and some are predominantly calcareous, while others are mainly siliceous. Pteropod and globigerina are the commonest calcareous muds, and radiolaria and diatoms are the commonest siliceous. There are also fine muds called **red clay**, which is the dust deposited from the atmosphere after a volcanic eruption.

Peléan Relating to Mt Pelée, a volcano in Martinique, which had a violent eruption in 1902. Mt Pelée was an explosive and very acidic volcano. It sent a cloud of gas (nuée ardente) down the mountainside after the explosion. The gas killed everyone in the nearby town of St Pierre, with the exception of one man, who was in a prison cell beneath ground level, awaiting execution. He was later rescued and reprieved.

peneplain* An old age landscape, such as Anglesey, in Wales and the Canadian Shield in North America. When mountains and plateaus have been eroded for millions of years, they are worn down until they are low and fairly flat. The more resistant hills may be left standing up above the general level of the plain, as residual hills or monadnocks. The peneplain will have been reduced almost to a plain by the effects of all the agents of denudation, but especially by the action of rivers. Anglesey is a small example of a peneplain, but a much larger example is the Canadian Shield in North America. A peneplain is the final stage in the cycle of erosion, which wears large mountain ranges down until they end up as low landscapes. See also **cycle of erosion**.

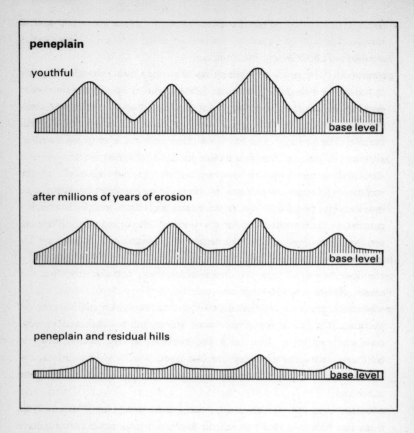

peneplain

youthful

base level

after millions of years of erosion

base level

peneplain and residual hills

base level

peninsula A narrow strip of land which protrudes into the sea. It will be attacked by marine erosion on both sides, and joints are likely to be enlarged to form caves. Peninsulas often have cliffs along one or both sides. Chalk, limestone, granite and sandstones all form peninsulas in different parts of Britain.

per capita income The income per person, generally quoted over a period of one year. The income of a country, or region, is divided by the number of people living in that country, to give an average income for every wage earner. The mean or average per capita income for a year is often used as an indication of wealth and economic development in a country. In many European countries the per capita income is several thousand dollars per annum, whereas in many African countries it is well below 1000 dollars per annum.

percolation The process by which water passes down through the pores in the soil and through cracks and joints in the rocks. The water moves downwards under the influence of gravity. The rate of percolation depends on the rock type, as well as on the amount of rainfall. Percolating water can carry dissolved chemicals, and may cause leaching.

percoline A line of water seepage through soil; part of the general throughflow movement. Percolines follow the general slope of the surface, and may have formed along lines created by worms or burrowing animals. There may be several percolines, which are roughly parallel, or they may have formed in a more dendritic pattern. They are situated just a few centimetres beneath the surface.

perennial irrigation see **irrigation**

perennial stream A stream or river which flows permanently. See also **exotic stream, intermittent stream.**

periglacial Around the edge or on the periphery of a glacial area. A periglacial landscape was not covered by ice, but was affected by cold conditions such as those now experienced in the tundra areas of northern Canada or northern Soviet Union. Southern England experienced periglacial conditions when central and northern England were covered by ice during the glacial phases of the ice age. Central and northern England, Wales and Scotland experienced periglacial conditions just before the ice covered the land, and for a period after the ice had melted. During the periglacial period there was permafrost, and frost heaving, freeze-thaw activity and solifluxion occurred. Some of the dry valleys of Britain were probably formed at this time. Other periglacial features include stone circles and stone stripes, which can be seen in the tundra areas of the Arctic. Also, there are accumulations of combe rock, which are the result of solifluxion.

periphery The edge or outside. The core-periphery idea is based on the studies of economic development, which suggest that the main towns, especially capital cities, are the core regions, where most money, industry and wealth can be found. The more isolated or peripheral areas tend to receive less investment and are not as developed. See also **core**.

permafrost Land that is permanently frozen, often to a considerable depth. The top few centimetres will generally thaw out in the summer, but the meltwater will not be able to sink into the ground because of the frozen subsoil. Drainage takes place only if the water can flow down a gradient. If the landscape is fairly flat, there will be surface water lying on the surface throughout the summer. This is found in many tundra areas, and the watery surfaces are excellent breeding grounds for mosquitoes and other insects. The insects attract many birds, which spend the summer in these areas, but migrate south for the winter. The wet surface layer causes serious problems to economic development, because the water freezes and thaws repeatedly as the temperatures change from day to night, causing the surface of the land to move and heave. Road building, pipe-laying, house construction, etc. become very difficult, and many different methods have been employed in Alaska, Canada and the Soviet Union to overcome the problems.

permanent pasture An area of grassland used for feeding cattle or sheep for many years.

permeable rock Any rock which is porous and absorbs water. Water can percolate through the pores down to the water-table to provide a supply of ground water. Most sandstones are permeable.

pervious rock Any rock through which water can pass by means of cracks and fissures. Carboniferous limestone and granite are good examples.

petrochemical industry A chemical industry which uses petroleum as its main raw material. A wide range of products can be made from petroleum, including many plastics. Petrochemical industries are often located near oil refineries; for example, Milford Haven and Teesside.

pH A measure of soil acidity, in which 7.0 is neutral, 6.0 or less is regarded as acidic and 8.0 is quite alkaline. It is a simple method of classifying soils, and is increasingly used by keen gardeners. Cereals grow best where the pH is about 6.5.

physical geography One of the two major parts of geography, the other being human geography. Human and physical studies are often closely linked because of the effects physical features can have on man, and the way that man can affect the landscape. Physical geography includes such subjects as biogeography, climatology, ecology, geomorphology, hydrology, meteorology, and pedology.

physical weathering see **weathering**

physiography see **geomorphology**

pie chart A circular diagram divided into segments which represent different percentages of a set of information. Also called **pie graph**.

piedmont Relating to the foot of a mountain. The plateau at the eastern margin of the Appalachians in the United States is called the Piedmont Plateau, and there is also a Piedmont (Piemonte) region in northern Italy. When two or more glaciers emerge from valleys onto a plain, and spread out until they merge with their neighbours, the resulting icefield is called a **piedmont glacier**, examples of which can be seen in Alaska.

pig iron The halfway stage between iron ore and the finished product, steel.

pike A summit of a hill in the Lake District and other locations in northern England. It is generally a rocky and fairly rugged summit.

pipeline A continuous length of piping for transporting liquids or gases. Pipelines are used to carry oil from many oilfields to ports or to refineries, and some natural gas is transported by pipeline for example, in the North Sea. Water is carried by pipes from reservoirs or rivers to houses and factories, and it may be piped to the fields in irrigated areas, in order to reduce the loss from evaporation, which is enormous if open canals are used.

placer A deposit of sand and gravel containing particles of a precious mineral, such as gold, tin, or platinum. River erosion further upstream will have created sediment which is then transported down the valley. The accumulated sands often contain particles of heavy minerals, which can be obtained by panning. There were gold rushes to Alaska and many other places, to search for deposits of this kind, and tin is still obtained from placer deposits in Malaysia.

plagioclase see **felspar**

plain An extensive area of low-lying, flat or gently undulating land. Some plains are the result of denudation wearing away higher land to form a peneplain, such as Anglesey or the Canadian Shield. Most plains are the result of deposition. Some of the flattest areas on earth are in regions where land has been reclaimed from the sea, as in the Fens, the Somerset Levels and the polders in the Netherlands. Other very flat areas are lacustrine plains, where former lakes have been filled in with sediment. Many valleys in the Lake District used to contain lakes which have dried up to leave very flat valley floors, as in the Great Langdale valley. The Vale of Pickering in North Yorkshire is also a lacustrine plain, and an even larger example is in the Red River valley to the south of Lake Winnipeg, in southern Canada and northern United States. Flood plains or alluvial plains are in river valleys, and glacial plains occur where large expanses of till, or glacial drift, have been deposited to cover the landscape, as in Norfolk, and eastern Denmark. Plains can be covered by any type of vegetation; for example, the Amazon lowlands are an alluvial plain covered by forests. There are several large plains which are covered by grasslands, and their names are sometimes used both for the type of vegetation as well as the geomorphological landscape feature; for example, prairies, steppes, pampas.

planetary winds The major winds of the world, which blow over very large areas. In tropical latitudes these are the trade winds, and they are very persistent. In the Northern Hemisphere the trades are north-easterly, and blow from areas such as the Sahara towards the equator. In the Southern Hemisphere they are south-easterly. Both the north-easterly and south-easterly trades blow from the tropical high pressure zones (the horse latitudes), towards the equatorial low pressure zone (the doldrums). On their journey they are affected by the rotation of the earth which deflects Northern Hemisphere winds to their right, and Southern Hemisphere winds to their left. In temperate latitudes the planetary winds are the westerlies, which are far more variable than the trade winds. In the Northern Hemisphere they are south-westerlies, and in the Southern Hemisphere they are north-westerlies known as the roaring forties.

plankton The very small plants and animals which float or drift around in the sea. Many of them are so small as to be invisible to the naked eye, and many are young plants or animals which will grow much bigger. They are the main source of food of many larger creatures, and are most abundant in the shallow areas of continental shelves, where sunlight can penetrate to provide food and energy. Plankton are often especially numerous in areas where cool and warm water meet. Such areas include the North Sea around Britain, the Grand Banks off Newfoundland and near the Humboldt current off Peru, which are among the most prolific fishing areas.

plantation A large farm or estate, especially in the tropical latitudes, where the commercial production of one crop takes place. Many plantations were set up and financed by foreign capital, with foreign companies organizing the production of such commodities as rubber, cocoa, bananas and sugar cane. In the past, many plantations used slave labour, notably the cotton plantations of the southern United State, but nowadays plantations often provide a good source of employment for local people. Most of the plantation crops are for export, especially to North America and western Europe, and, as much of the plantation farming is organized by large international companies, many of the profits also go to North America and western Europe. Plantations often take up good farm land, which could be used for growing food crops for the local people. Positive features of the plantations are that, in addition to providing employment, they have often provided schools and hospitals, paid for housing for the workers and probably contributed to the construction of roads in the area. Other aspects of infrastructure, such as water supply or electricity, may also be developed by the plantation company. Major plantation areas include coffee on the Brazilian plateau, rubber in Malaysia, bananas in the West Indies and Central America, cocoa in Ghana and Nigeria, and tea in India, Sri Lanka and China.

plateau An elevated area of land which is fairly flat. The plateau may be broken by deep river valleys, making it a **dissected plateau** such as the Meseta of Spain. Mountain ranges often stand up above the general level of the plateau, as on the Meseta or the plateau of Brazil. Some plateaus may be very high, over 3000 m above sea level, as in Tibet. Even though they are high, plateaus can be surrounded by mountains, and are then called intermontaine plateaus; there are examples in Peru,

Ecuador and Bolivia in the Andes. Some plateaus may be the result of uplift between faults to form block mountains, such as the Massif Central or the Vosges in France. The plateaus of southwest United States consist of almost horizontal strata; this has contributed to the formation of large canyons, which dissect the plateaus.

plate tectonics* The study of the formation of the major structures on the earth's surface by the movement of the underlying plates. If they move together, the colliding plates may cause earthquakes, volcanoes and folding, but if they move apart, basaltic lava eruptions are the most likely result. The old theory of drifting continents is now related to the movement of the plates, some of which carry continents as though they were passengers on a raft. Where plates are diverging, that is, moving apart, as along the mid-Atlantic ridge, especially in Iceland, there are frequent extrusions of magma to form new basaltic rocks. Sea-floor spreading is occurring in these locations. At convergent boundaries one plate sinks below the other in the subduction zone; this process is happening along the western edge of the Pacific. The plates are affected by the convection currents in the earth's mantle, and really belong to the field of geology. However, because of their influence on mountain formation, they are also a very important part of geography.

Pleistocene A geological period, which takes it name from Greek words meaning most recent. It can be used to refer to the last two million years including, or excluding, the most recent time since the last ice advance. The post-glacial period is called the Holocene. The beginning of the Pleistocene is now generally agreed to be about two million years ago; in some places it coincides with the beginning of the ice age, but as the ice advanced over different places at different times, there is no set time for the beginning of the glacial phase. In Britain, the Pleistocene consists of glacial and interglacial phases.

plucking A process associated with glacial conditions. When ice in a glacier freezes to a rock on the valley floor or side, the movement of the ice is likely to pull part of the rock away. Plucking gives rise to jagged rock surfaces. It is an important process on the back walls of corries, and also occurs on the down–valley side of a roche moutonnée.

plug see **volcanic plug**

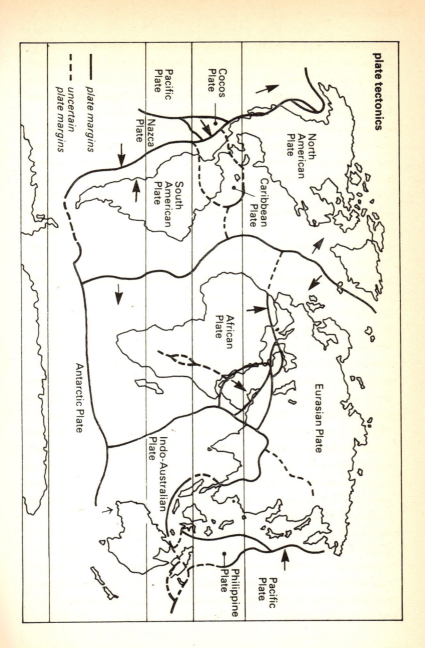

plate tectonics

North American Plate

Cocos Plate

Pacific Plate

Nazca Plate

South American Plate

Caribbean Plate

plate margins
uncertain plate margins

African Plate

Eurasian Plate

Antarctic Plate

Indo-Australian Plate

Philippine Plate

Pacific Plate

plutonic rock An igneous rock which has formed beneath the surface. Molten magma will have cooled slowly, and there will have been time for large crystals to form. All plutonic rocks are crystalline and coarse grained. The commonest plutonic igneous rock is granite, which is light in colour because its mineral content is acidic. The commonest of the basic plutonic rocks is gabbro, which can be seen in the Cuillin Hills of Skye and in the Lizard Peninsula. Most plutonic rocks will have been formed in batholiths beneath the surface, but because of the erosion of overlying rocks, they are now commonly found on the surface. The granite masses of southwest England, Dartmoor, Bodmin Moor, Hensbarrow, Carnmenellis, Land's End and the Scilly Isles are all plutonic.

podsol Also **podzol** The commonest type of acidic soil, often occurring in areas of coniferous forest, where the needles fall to the ground and add to the acidity. The word podsol is Russian in origin and means ashy grey, which is a common colour of podsols. They are leached soils, and the iron and lime which have been leached from the A horizon are often deposited as a hardpan layer in the B horizon. Large areas of podsol are found in the taiga regions of northern Canada, Finland, Russia and Siberia. There are also smaller areas of podsol in Britain's coniferous woodlands, and in some heathland areas in Surrey, Hampshire, and elsewhere. They are normally quite poor agricultural soils, but can be improved by the addition of fertilizer.

polar front The meeting point between cold polar air moving southwards and warm tropical air heading northwards. It is commonly located in the western part of the Atlantic near to the northern United States. As the two air masses pass each other, a whirling mass of air can develop, and in this way depressions are initiated. Once a depression has fully developed, the polar front will have been swamped by the whirling mass of air, which will then begin to travel eastwards across the Atlantic.

polar winds Cold air which blows from the Arctic or Antarctic. The high pressure region near the North Pole is a source of outgoing winds. They form large air masses, which travel southwards. A polar air mass moving towards Britain will bring cold air, though it will warm up slowly as it travels across the North Atlantic. As it warms up, it gains moisture from the ocean, so that snow showers may develop in the winter months, and heavy falls may affect the hills of Scotland, especially in north-facing locations.

polder An area of flat low-lying land which has been reclaimed from the sea, especially in the Netherlands. The land is surrounded by embankments, called dykes, to keep the sea out, as many polders are below sea-level. Polders need draining by pumping the water out into the ditches, which are numerous in the low-lying parts of the Netherlands. They are very fertile and can be used for growing cereals or bulbs, though much of the polderland is still left as rich pasture for feeding dairy cattle. There are small areas of polders in neighbouring parts of Belgium and West Germany.

pole 1 Each of the extremities of the earth's axis; they remain stationary whilst all other parts of the earth are rotating around the axis. 2 See **growth pole**.

pollution Contamination of the atmosphere or water, especially by chemical or industrial waste. Air pollution affects the atmosphere, and subsequently the land and oceans. Water pollution affects rivers, lakes and the sea. In the past it was quite common for industries and urban areas to use rivers and the sea as dumping grounds for their waste. There is now so much pollution that it is no longer possible to get rid of waste in this way, and it is known that some of the pollutant material is positively harmful. Much environmental damage has been done in the past through ignorance or carelessness, and there are now increasingly strict controls to try to prevent further damage. Clean air acts have reduced the amount of smoke, and in London, for example, smogs have virtually disappeared. They still occur in some locations, such as Los Angeles, where motor cars are the cause of the pollution. There is also a great deal of pollution in the atmosphere, and power stations seem to be a major cause. Most European countries are now cleaning up their thermal power stations in order to reduce atmospheric pollution. It is believed that acid rain has been caused by industrial pollutants which have been emitted into the atmosphere and then brought down to earth in rain. Many lakes have lost their fish; millions of coniferous trees have died, and it is thought that pollution has been the cause. Even stricter controls are necessary, and most developed nations are deeply concerned about the seriousness of the problem. Third World countries have less pollution at present because they have less industry, though in some localized areas they too have serious pollution problems. One of the most seriously contaminated and poisoned areas is at Cubatão near São Paulo in Brazil. There is increasing anxiety about pollution of water supplies in western Europe and North America because of agricultural fertilizers which have been washed into streams, where they can be harmful to animal life and possibly to humans. There is still much to be learnt about the harmful effects of chemicals used by farmers, as well as by industrialists.

population 1 The total number of inhabitants of a city, country, or region. In a United Kingdom census the total population is surveyed every ten years, but a 10% sample is studied in the fifth year between two major censuses. 2 (in statistics) The total number of anything, whether people, cities or trees in a park.

population change The increase or decrease of population in a country, region or city. The major reasons for change are variations in the birth and death rates, and in some areas migration will be important.

population density The number of people per square kilometre. The figure can vary from 1 or less in parts of interior Australia or the Amazon basin, to over 1000 in parts of Singapore and Hong Kong. The average figure for the United Kingdom is 228 per sq km, and for Greater London the figure is 500. The last figure is very similar to the figures for many other large cities, including New York and Chicago.

population distribution The way in which a population is dispersed. It can be considered on a variety of scales. The world as a whole has many densely populated areas, notably parts of China and Japan, much of Java, many parts of western Europe and northeast United States. The world's sparsely populated areas include Australia, Canada, the Sahara and Antarctica. On a smaller scale, in Europe there are many well-populated areas, including southeast England, and there are sparsely populated areas, such as northern Norway and much of northern Scotland. On a still smaller scale, in southeast England there are sparsely populated areas of heathland and woodland even in the commuter belt of Greater London.

population explosion A large and rapid increase in the size of a population. With the help of medicines, chemical sprays and money from more advanced countries many of the Third World countries experienced a dramatic reduction in their death rates during the 1950s, but there was no accompanying reduction in the birth rate. Previously the high birth rate had been matched by a high death rate, which ensured that population growth was quite slow. With this sudden change in circumstances, the population grew rapidly in what is now described as a population explosion. In the more advanced countries of western Europe the birth and death rates had been changing gradually over the previous hundred years or so, but the Third World went through a drop in death rates from 40 per thousand to less than 20 per thousand in only a few years. The corresponding change in birth rate comes more slowly with education and improving economic standards, and some countries have not managed to achieve much progress. Birth rates in Kenya for instance are still 56 per thousand, compared with the figure of 12 for the United Kingdom. The rapid increase of world population during the 1960s and 1970s was sometimes said to be exponential, and shows up very clearly on world population graphs, and in the demographic transition.

population growth An increase in the population of a city, county, or country. The growth is the difference between births and deaths, with the added influence of migration in some cases. The world population has grown at a varying rate in the past, and the average at present is 2% per annum; in the 1970s a figure of 3% was normal. The world's population had grown to one thousand million by 1820, and reached two thousand million by 1930. Three thousand million was reached by 1960 and four thousand million by 1975. Total population is now five thousand million, but the rate is slowing down and may not reach ten thousand million by the year 2000, although that was the forecast in the early 1980s.

population pyramid see **age sex pyramid**

porous The terms used to denote rocks containing pores (very small holes), which retain water. Sandstones are from 5 to 15% pore space, and loose gravels are up to 45% pore space. Most porous rocks are permeable, except for clay, which has such small pore spaces, that they are blocked or sealed by water held in place by surface tension.

position The location of a settlement, in general terms, as opposed to specific detail. The original site of London was on an area of slightly higher and drier ground in a predominantly marshy plain near the Thames. Its position is in southeast England, at the lowest bridging point of the Thames. Also called **situation**.

post-industrial A term used to denote any of the more advanced nations which have seen a decline in the old heavy industries, notably steel, but also shipbuilding and often textiles. The older industrial nations have developed many new industries, mainly connected with electronics, computers, or automation, and have a very large service sector of industry. Britain, the United States and West Germany are good examples.

pothole 1 A small, roughly circular hole on the rocky bed of a stream. Only a few centimetres in depth and diameter, it will be the result of corrasion, as small pebbles have been whirled round and scraped on the bed and sides of a small indentation. 2 A larger, vertical hole in an area of limestone, occurring where a joint has been enlarged by solution. If a pothole is used by a stream it is called a swallow hole. Potholes generally extend from one bedding plane to the next and along the bedding planes there may be caverns or channels eroded by flowing water. Potholers can progress up and down using potholes, and move horizontally by means of the river-formed channels. Many of the channels have become dry because the river has changed its route, probably to a lower level. Potholes in limestone are particularly well developed in the Ingleborough and Malham areas of the Pennines and near Cheddar in the Mendips.

prairie A flat grassland area. The prairies are the mid-latitude temperate grasslands of central United States and Canada, and extend from Alberta, Saskatchewan and Manitoba in Canada, southwards through the Dakotas, to Kansas, Oklahoma and Texas. They are flat and treeless areas, and the climate is ideal for grass to grow. There is summer rainfall, 700 mm per annum in the east but decreasing westwards to only 300 mm, and the winters are fairly dry, with just a little snow in the north. Summer temperatures are 20°C or more, but winters can be very cold in the northern parts; temperatures of $-10°$ to $-20°C$ are often experienced. Large areas of the prairies have been ploughed for cereal production, and the wetter eastern parts are very productive. The drier western prairies, where the grass is shorter, are less productive, and repeated ploughing in several places has given rise to dust bowl conditions. Before the ploughing took place, the grasslands were the home of large numbers of cattle and sheep, which were looked after by cowboys. Prior to the European settlement, the prairie landscape had been the home of millions of buffalo and numerous groups of Indians. The steppes of the Soviet Union are very similar to the prairies, and in the Southern Hemisphere there are also temperate grasslands: the pampas, the veld, the Canterbury Plains and the Murray Darling Basin, but they are milder and wetter than those of the Northern hemisphere.

Pre-Cambrian The earliest geological period, which finished about 600 million years ago. Nearly one quarter of the world's land surface is covered by Pre-Cambrian rocks, but very little is known about them because they have been eroded, folded and faulted so much, and most have been metamorphosed. Also, there is very little fossil evidence. Because only very primitive forms of life existed in those days, it is more difficult to work out the palaeogeographic conditions of the time than in the case of more recent rocks. Pre-Cambrian rocks occur over large parts of southern Africa, much of Canada, Western Australia and large areas of Siberia. The Baltic Shield in Finland and neighbouring countries is the largest European area of pre-Cambrian rocks. The largest area in Britain is in northwest Scotland, where the Lewisian gneiss occurs, and there are smaller outcrops in Anglesey, the Malverns, Wrekin, Charnwood Forest, and elsewhere.

precipitation The products of condensation in the atmosphere which fall on the surface of the earth, including rain, snow, hail, sleet, dew and hoar frost. They can be measured in a rain gauge, although the snow and frost would have to be melted first. The total precipitation is recorded in millimetres.

pre-industrial A term used to denote an early, mainly rural society in which agriculture was the predominant activity. Pre-industrial societies were originally subsistence, but gradually trade developed as communications improved. The use of metals and the development of craft industries contributed to the growth of towns and the development of non-agricultural settlements. Industrialization began with the use of steam power at the end of the 18th century, after which many towns grew rapidly, and an urban society evolved. Industrialization took place at different times in different countries, as outlined in Rostow's model.

pressure, atmospheric The weight exerted by a column of air on the earth's surface at any given point, normally at sea level 14.70 lb/in^2 or 1033.3 g/cm^2. Pressure is normally measured in millibars. The lowest sea level pressure recorded in Britain was 925 mb and lowest in the world was 877 mb. The highest ever recorded in Britain was 1055 and in the world 1079 mb.

pressure gradient The change of pressure as shown by isobars on a weather map. The isobars are similar to contours, and the pressure gradient can be compared to a hill. The closer the isobars, the steeper the hill, and the stronger the wind. Few isobars will indicate weak winds.

prevailing wind The most frequent wind direction. Some parts of the world have very persistent winds; for example the trade winds. In temperate latitudes the winds are more variable, especially where depressions occur. Britain has prevailing westerly weather, because so much of the weather comes from the Atlantic in the form of depressions. In depressions the winds blow from all directions, and so they do not bring dominant westerly winds. Over the period of a year, the commonest wind direction in Britain will be south-westerly, but there will be many winds from south, west, and other points of the compass. Some parts of eastern England receive more winds from the east than from any other direction in the winter part of the year.

primary era see **Palaeozoic**

primary industry Any of the industries which provide primary products, that is, goods that have not been changed or processed. The primary industries include mining and quarrying, forestry, agriculture and fishing, and together they consititute the primary sector of the economy.

primary sector The part of the economy which is concerned with the primary industries. Processing and manufacturing is regarded as the secondary sector; the tertiary includes all the services, while quaternary industry is the information and expertise sector. The percentage of working people employed in each of these sectors is a guide to the stage of economic development reached by a country. Third World countries are still very much involved in agriculture, in the same way that Britain and other European countries were 300 years ago. Only 2% of the working population of Britain is now involved in primary activities, compared with over 90% in some countries, such as Mali or Nepal. In addition to national differences, there are also regional differences in employment. For example, southeast England has a higher proportion of people employed in tertiary and quaternary activities than northwest England.

primate city The largest city in a region or country, often the capital city, such as London or Paris. In some countries, such as the United States or Brazil, the capital is not the largest city. London and Paris are much larger than the second cities in size in England and France, and they are said to have a high degree of primacy.

primeur (*French*) A term denoting early fruit and vegetables. If early produce can be taken to customers before rival producers have grown their fruit and vegetables, a higher price will be obtained. In France, the southern Mediterranean areas are able to grow tomatoes, strawberries and flowers for Paris and other markets much earlier than they can be produced in the north of France. In the United Kingdom, the Scilly Isles are able to grow daffodils earlier than anywhere else, Cornwall grows early potatoes, and the Channel Islands grow fruit, flowers and vegetables. In the United States, it is Florida and California that control the early season crops. Nowadays early produce can be grown in greenhouses, using artificial heating, but this makes it more expensive. In Iceland greenhouses grow bananas, peppers and other products using heating provided by hot water from underground.

processing industry An industry which converts an agricultural commodity into a saleable product. Examples include obtaining oil from olives, juice from oranges, cheese from milk and wine from grapes. Processing may involve purification or refining.

profile* A vertical section of soil, which shows the changes in the soil, including the different horizons from the surface down. **2** A drawing of a section of a river valley. See also **river profile**.

promontory A headland or small peninsula which protrudes into the sea. It will suffer from erosion along both sides, and is likely to have cliffs and possibly caves as a result of wave action.

pull factor see **migration**

pumice A type of volcanic rock which is very light and porous. It contains a large number of holes, which are the result of escaping steam and gases at the time of its formation. It is so light it will float, and it tends to be very acidic in chemical content. Much of it originates as a kind of froth or scum floating around on the top of lava.

pumped storage A system employed at some hydro-electricity stations, whereby water is pumped back up hill to a storage reservoir during the night, when surplus electricity is available. The water can be reused the next day to generate electricity, and it helps power stations to operate with quite small storage reservoirs. The first example in Britain was at Cruachan in Scotland, but there are larger examples at Ffestiniog and Dinorwic in north Wales.

puna An area of high, level and bleak land in the Andes, between 3000 m and 4000 m above sea level. Temperatures can be low, even in mid-summer, as the atmosphere is rarified because of the altitude. Summer days may reach 10°C, but the night temperatures fall well below freezing point. Throughout the year, bitterly cold winds are common. The local Indians sometimes grow hardy cereals and potatoes, and rear llamas and alpacas. Life is very hard and the people are poor. They keep warm by staying in the sun for as long as possible, but they wear woollen ponchoes made from their own animals. There are a few rich mineral deposits on the puna of both Peru and Bolivia, and the local people provide a source of low-paid labour, the profits going mostly to a few rich landowners.

push factor see **migration**

puy A volcanic plug, taking its name from Le Puy, a town in the Massif

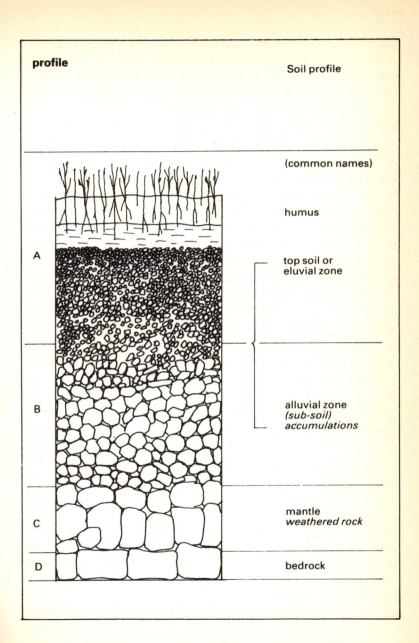

profile

Soil profile

(common names)

humus

A — top soil or
eluvial zone

alluvial zone
(sub-soil)
accumulations

B

C — mantle
weathered rock

D — bedrock

Central in France, where there are several examples, including one which has a monastery built on its narrow summit.

Pygmy   A race of small people, who grow to about 135 to 140 cm. The most famous groups are the negrillos of the central African forests, but there are others in southeast Asia, who are known as negritos. They are mostly semi-nomadic hunters and gatherers who live in tropical forest areas. They are decreasing in numbers because of outside influences, and many of them have become settled farmers or shifting cultivators.

pyramid peak   see **horn**

pyrites   Any of various yellow minerals with a metallic lustre, especially iron pyrites (FeS_2), which is a source of iron and sulphur. Sometimes its colour has led to its being mistaken for gold, and it is often called fool's gold. Copper pyrites is similar to iron pyrites, but has a different mineral content. It is found in some igneous rocks, and often occurs in veins or seams of igneous materials which have forced their way through existing rocks in liquid or gaseous form. When they solidify (for example, in the limestone areas of Derbyshire), they may be mined or quarried for a variety of minerals, not only pyrites.

pyroclastic rock   A fragment of rock thrown out by volcanic activity, generally in association with some violent explosive activity. Pyroclasts normally include solidified lava left behind by a previous eruption of the volcano, as well as rocks from the crust, smaller pieces of cinders, ash and dust. The largest pyroclastic fragments may weigh several tonnes, and they are called **volcanic bombs**. Smaller pieces are called **lapilli**. When the pyroclastic activity ceases, there often follows outpourings of lava.

Q

qanat Also **kanat, karez** (*Persian*) An underground irrigation channel which brings water from foothills to the neighbouring plains. There are examples in Pakistan and Iran. Also called **karez**.

quadrat An area of land, generally square in shape, in which a field study is taking place. Quadrats are used most frequently for vegetation surveys in ecological studies; the most popular size is 1 sq m. Within the area of the quadrat a detailed study of the plants can be made; it will be regarded as a sample study, which can be compared with other quadrats studied elsewhere. On a transect line across a valley, for example, it would be possible to take a quadrat sample of vegetation every 10 or 100 m, depending on the scale of study. If trees were the major subject of the study, a quadrat of 10 or 20 m would be more suitable.

quality of life The standard of living together with other, less definable considerations, such as the amount of noise and pollution, the availability of countryside, flowers and birdsong, etc. Conservationists in particular are very concerned about such considerations, which are not always available to people living in large towns, although they will have museums, theatres, etc., which are also factors to be considered. Geographers have to try to consider all aspects when assessing problems such as conflict over land use.

quantitative Considering quantities. Geographical studies became increasingly quantitative during the 1960s and 1970s, and the purely descriptive qualitative work became less important, although still an essential part of geographical studies. Instead of describing slopes, for example, geographers went out and measured them and then tried to explain how they had formed and why they were different from other slopes. Quantification meant that problems were solved by examining data and trying to be objective rather than subjective.

quarrying The excavation of rock from an open working on the surface. Many building materials such as limestone, sandstones and granite, are quarried. Chalk is also quarried for use in cement and lime. Occasionally coal and iron ore are extracted in this way from open-cast workings. Quarries sometimes create conservation problems, as they can become eyesores. Many old quarries cannot be changed now, but in modern workings great care has to be taken of the landscape, and whenever possible trees are planted to hide the workings. Such care is not always possible, however, and the Peak District National Park has a particular problem, as essential supplies of limestone are being quarried inside its boundaries.

quartz A very common mineral, found in many of the earth's rocks. It consists of silica; that is, silicon and oxygen. Most grains of sand consist largely of quartz, and it is very common in granites and other acidic igneous rocks. Pure quartz is colourless and glassy, and it is the major source of glass. It is a hard mineral, the hardest of all the common substances.

quartzite Metamorphosed sandstone which is very hard and resistant. It consists mainly of quartz and is cemented by quartz. Old quartzites are found in the Cambrian rocks of the Shropshire–Wales border region, in Hartshill Quarry in the Midlands, and near Durness in northwest Scotland. There are some softer quartzites, which are sedimentary in origin.

Quaternary period A geological time period representing the last two million years. It consists of the Pleistocene, which was the ice age, and the Holocene which is post-ice age. During the Quaternary there were several ice advances, followed by retreats during which interglacial conditions prevailed. These were sometimes warmer, but sometimes cooler than the present climatic period. The present is thought likely to be an interglacial period, rather than definitely postglacial. Glacial periods were similar to the conditions of Greenland and Antarctica at the present time. Interglacials included some tundra conditions with periglacial processes being active, as well as some warmer periods when deciduous forests were able to grow.

quaternary sector The part of the economy which deals with providing information and expertise. The quaternary sector has grown rapidly in recent years. Quaternary sector industries are found mostly in advanced countries, especially near main centres of modern industrial development, for example, near London and the Thames valley–M4 corridor.

quicksand An area of sand and mud which contains a high proportion of water. It generally occurs near a river or the sea, where the water-table is high, but drainage is poor. It is because quicksand is almost liquid that animals may sink into it.

R

radial drainage A pattern of rivers in which the flow of water is away from a central high point. The rivers all radiate outwards, like the spokes of a wheel. The central high point is likely to have been a dome originally, although erosion over millions of years will have changed its shape. The Lake District is a good example, and the radial pattern has been accentuated by the formation of lakes during the ice age.

radiation The emission of energy in the form of heat. Radiant energy is given off by the sun, and some reaches the earth, where it provides heat. The earth gains heat from radiation from the sun, but also loses heat by radiation from the ground. During the day, more heat is received from the sun, by insolation, than is lost by the ground. At night, the loss is much greater. There are large differences between the seasons, because the long days in summer enable the ground to warm up, whereas in winter it will become progressively colder. Land loses and gains heat more rapidly than water, which contributes to the climatic differences between the continents and the oceans.

radiation fog Fog which occurs as a result of heat loss from the ground by radiation during the night. Clear skies and calm weather conditions are ideal for the formation of radiation fog. See also **fog**.

radioactive decay The breakdown of certain unstable particles in rocks which enables the age of the rocks to be calculated. The rate of change is based on half-life method of dating. This means that half the radioactivity is lost after a set time – say 20,000 years. After a further 20,000 years there will be another loss of 50% and so on, since there is a gradual decrease in radioactive content. Each mineral has its own half-life period. Uranium^{238} changes to lead^{206}; uranium^{235} to lead^{207}; thorium^{232} to lead^{208}.

radiocarbon dating A method of dating using the radioactive isotope of carbon called C^{14}. Carbon 14 oxidizes to carbon dioxide. It enters the earth's carbon cycle and is absorbed by plants. Once the plants have died, the carbon 14 diminishes at a known rate. The rate of decay is known, because carbon 14 has a half life of 5570 years. The age of wood or peat, as well as bones or shells, can be dated by analysing the proportion of radiocarbon in the total amount of carbon contained. The radiocarbon method can be used for measuring ages of up to about 70,000 years, although above 25,000 to 30,000 years it becomes less accurate.

rain Drops of water falling from the sky. They are formed by condensation of water-vapour in the atmosphere, caused by cooling of the air. Small droplets form cloud, but as the drops grow they become too heavy to float, and so they descend to the ground. They generally have to exceed 0.5 mm in diameter in order to overcome the effects of slowly rising movements of air.

rain day A period of 24 hours in which a measurable amount of rain has fallen. The minimum measurable amount is generally taken to be 0.25 mm.

rainfall 1 A shower or fall of rain. All rain is formed by rising air, which is then cooled to enable condensation to take place. The air rises for a variety of reasons and so there are different types of rainfall. They are convectional, cyclonic, frontal and orographic. 2 The amount of precipitation received by a given area over a given period.

rain forest A dense forest growing in regions with heavy rainfall throughout the year. Such forests are most numerous in tropical latitudes, such as the Amazon Basin, Zaire, and parts of southeast Asia. There are examples in warm temperate regions such as eastern Brazil, Natal in southern Africa, and parts of China and southern Japan. There are also examples in wet maritime regions, such as Tasmania, southern New Zealand and southern Chile.

rain gauge An instrument used for measuring rainfall and other forms of precipitation. There are many different types, including some self-recording varieties. They consist of a cylinder which stands up above the ground surface to prevent splashing water bouncing in. Inside the cylinder there is a funnel to lead the water into a collecting jar. The water collected is poured into a measuring cylinder, and the total rainfall can be measured off, generally in millimetres. The commonest rain gauges have a diameter of about 13 cm, and will be read every 24 hours.

rain shadow see **orographic rainfall**

rain wash A process in which drops of rainwater hit rock and soil, washing away tiny particles. It is especially effective in areas with no vegetation cover, and can be a cause of soil erosion. Rain wash is a major contributor to the erosion of valley sides, and helps to create the V-shaped cross-profile of valleys in humid regions. The lack of rain wash in deserts is a reason for the steep sides of wadis and canyons.

raised beach An elevated beach, which is now above sea-level because of an uplift of the land or a retreat of the sea. It is often bounded on its inland edge by old cliffs, sometimes with caves. The uplift of land may have been the result of tectonic movements, or may be the result of isostatic readjustment of the land after ice has melted. When the weight of ice has been removed by melting, the land gradually moves up, relative to the sea. Raised beaches of up to 60 m have been formed in Scotland because of isostatic movement, and some of over 200 m are found in Norway. The same effect can be produced by the sea's retreating, and this happened when snow and ice increased during the glacial phases. More snow and ice meant less water in the sea, and so the sea-level went down. Some locations have a series of two or three raised beaches because of the several changes of sea level during the ice age.

rakes see under **lode**

ranching A type of pastoral farming, generally found in grassland areas. Large scale and extensive methods are used to rear animals, usually cattle, but also sheep. Formerly, the animals could roam around over large areas and were looked after by cowboys, but nowadays there are more fences to restrict the movements of the animals. Cowboys still exist in some places such as Northern Territory in Australia, but motorbikes are often used instead of horses. The western United States, Argentina, Uruguay and parts of Australia are major areas for ranching. Many of the animals are fed on alfalfa, corn and other feedstuffs, as well as fending for themselves on the natural pasture.

Randstad An urban area in the Netherlands which includes Amsterdam,

Rotterdam, The Hague, Dordrecht and Utrecht. It is a ring of almost continuous urban development, with a green belt in the middle of the ring. It is the original and best example of a **ring city**.

range 1 A line of mountains which makes a more or less continuous barrier, even though there will be passes or gaps through it. The Pennine Range is the largest in England, but the Appalachians or Rockies in the United States are much bigger. 2 An open area of grazing, as found in the high plains of the western United States. 3 The difference between the highest and lowest temperatures experienced in any given period, such as a day or a year. 4 The difference between the high and low points of a tide. The range will be greater at the time of spring tides than at neap tides. 5 The distance over which a commodity will be distributed from a central place. This depends partly on the maximum distance a consumer is willing to travel in order to purchase the commodity, which in turn depends on how frequently it is required. For instance, low order goods with a daily requirement, such as bread or newspapers, will be obtained from somewhere close by. High order and more specialized goods which are only required every few weeks or months may be obtained from many miles away.

ranking The placing of a set of data in numerical order so that the highest figure is ranked one, the next highest is ranked two, and so on. Ranked information is used in many statistical techniques, especially for correlation.

rank-size rule The theory that the population of the second largest city in a region or country will be half the population of the largest city of that region or country. The population of the third largest city will be one third that of the largest city, and the fourth city will have a population one quarter the size of the largest city. The formula employed in the rank size rule is $P_n = P_1 (n)^{-1}$, where P_n is the population size of a town of rank n, and P_1 is the population of the largest town in the area. The rank-size rule fits many countries, especially if industrialization has taken place. In some countries, such as Britain or France, the excessive size of the primate city makes all other cities look incorrect according to the rank-size rule. However, it is really the large size of London and Paris which upsets the theory in these two countries.

rapids Small waterfalls and turbulent water along a stretch of river. They are generally the result of a different and harder type of rock outcropping on the bed of a river. Rapids are gradually worn down by erosion, and eventually disappear. Steep rapids, such as those in the Nile, are called **cataracts**.

raw materials A primary or partly processed product which is the basis for manufacturing into a finished product. Raw materials are the necessary requirements of industries. Examples include agricultural products, minerals and timber as well as many half-made goods such as cotton which has been spun but not woven, wood pulp, or flour.

reafforestation The planting of trees in an area which has been deforested. In many parts of the world, trees have been cut down, and the land has been used for agriculture, as in Britain in the past and as is happening in Brazil today. In other places such as Nepal, trees are cut down on the slopes of the Himalayas, and immediately the exposed soil is washed away. It is very important that reafforestation should take place in such areas in order to prevent soil erosion. Other forested areas, such as Sweden and Canada, cut down large numbers of trees for timber and pulp, but they now plant as many trees as they cut down. Reafforestation is important in these areas too.

recessional moraine see **moraine**

reclaimed land Land which is recovered for productive use. It may be reclaimed from the sea, from marshes, or from dereliction after spoiling by mining, quarrying, or industry. Land reclaimed from the sea is called polderland in the Netherlands, and the Fens are the best example in Britain. Barriers called embankments or dykes are built to keep out the sea water, and pumping is necessary to keep the land dry enough for crops to be grown. Often grassland is grown on the reclaimed land and cattle can be reared, but in much of the Fens cereals and potatoes are grown. In Languedoc in southern France the land was marshy and malarial until the 1950s, but has now been reclaimed for tourist developments. Marshes near Naples in southern Italy and Catania in Sicily have been reclaimed, and are used for growing soft fruit and vegetables. There are many examples of land reclaimed from old industrial areas in the coal-mining regions of south Wales, central

Scotland, northeast England, northeast France, Belgium and West Germany. An outstanding area is the Swansea valley in south Wales, which won awards for turning a decayed and derelict area into an attractive parkland.

recreation A pastime, diversion or other leisure activity. Recreation is a growth area, both for holiday-makers who are away from home, and for local people who may be looking for recreation in a country park, funfare, theatre, or sports centre. Resorts which cater for sun-seekers or skiiers are a major branch of the recreation industry which has grown rapidly in the last 30 years, with the development of air travel, package tours, longer holidays and more money. The provision of recreation facilities in Britain has been helped by the development of the national parks, country parks and long distance footpaths, as well as many other attractions.

recycling The process of collecting and recovering waste for reuse. Many minerals can be collected as scrap, melted down and reused. Recycling is already commonly practised in steel manufacture, but can also be applied to lead, copper and tin. There are aluminium collecting centres in some places, but most forms of reusing metals are uneconomic at present. However, in order to conserve the diminishing supplies of metals, it will be vital for recycling to become more common. Recycling of bottles is also growing in importance, and there are several places where paper is recycled. Technology has been slow to develop equipment for reusing many of the raw materials which are exhaustible, because it has always been expensive, and there was no great need in the past. Now that it is realized that many of the earth's resources will be exhausted during the 21st century, there is increasing pressure on industry to reuse materials and not be as extravagant as in the past. This will inevitably add to the cost of many items.

red clay see **pelagic**

reentrant A small valley or lowland area extending into a region of higher ground. It will have been formed by water erosion or by freeze-thaw and solifluxion.

refining The process of removing impurities from metals, sugar, petroleum, etc. For metal ores especially, it is best to carry out the refining process near the mines in order to avoid unnecessary transport costs. Iron ore is from 25 to 70% pure, and a low-quality ore of only 25% contains 75 tons of unwanted rock in every 100 tons extracted. Whenever possible, unwanted minerals are separated out before transportation.

regelation The process of refreezing. In a glacier a considerable amount of melting occurs, because of the effects of pressure. The melting point of ice is lowered by the pressure caused by the weight of the ice. If the pressure is released, then the melting point will rise, and the water will freeze again. Regelation is probably very important in explaining the movement of a glacier.

regime The seasonal changes in climatic conditions. The way in which temperatures and rainfall change from month to month. See also **river regime**.

region An area which has a number of distinctive characteristics. It may be very small, such as a tiny valley, or very large, such as the Amazon basin. A region may be defined by one or several of its characteristics. Climate, relief, rock type, agricultural pattern, natural vegetation, and many other features may be the reason for the delimitation of a region. The Weald is a region because of its geomorphological characteristics, and Dartmoor is a region because of its rock type. The Sahara is a region because of its climate, and London is a region because of its urban development and buildings.

regolith Loose rock fragments which have broken away from the solid bedrock. The fragments may be very thin and are sometimes patchy. Eventually, they will be broken down into soil particles, and sometimes the word regolith is considered to include soil and superficial deposits such as alluvium, glacial drift and loess.

Reilly's Law of Retail Gravitation A model devised by the American, W.J. Reilly to enable predictions to be made about shopping habits and the likely limits of catchment areas for shopping centres. It uses the relationships between distance and population to provide the answers, and assumes that all people are logical in making their decisions about where to shop.

rejuvenation A renewal of the erosive power of a river because of a change of gradient in the river's long profile or the arrival of extra water. Extra water may be the result of river capture or because of melt-water from a glacial area, as would have happened at the end of each glacial phase in the ice age. A change of gradient would be the result of a fall in sea-level, or of a rise in land level near the source of the river. In any of these situations there would be renewed vertical corrasion along the river bed. Several distinctive features are formed by rejuvenation. The knick point is the point at which increased erosion begins, and is simply a break in the long profile of the river, marked by turbulent water or a small waterfall. The knick point is gradually eroded and the break on the long profile will work its way upstream. Rejuvenation causes increased erosion, and some meanders may become incised, as on the Wear at Durham and the Dee at Llangollen. The river may cut down to a new and lower level, leaving the old flood plain abandoned at a higher level as a river terrace.

relative humidity The amount of water vapour actually present in the air, expressed as a percentage of the maximum that could be held by air at a particular temperature. It is a measure of dampness, and is normally measured by a hygrometer using wet and dry bulb thermo-meters. The warmer the air, the more water vapour it can hold. It is possible for air on a dry day in the Sahara to be holding more water vapour than the air in Britain when it is actually raining. If air is cooled, it can hold less water vapour; for this reason dew is common during the night in the Sahara. It is also the reason for the formation of cloud and rain when air rises. Air at a temperature of 10°C can hold 9.4 gm per cu m; air at a temperature of 20°C can hold 17.3 gm per cu m; air at a temperature of 30°C can hold 30.3 gm per cu m.

relief The features of the landscape. Differences in altitude, steep slopes and gentle slopes, flat plains and cliffs are all part of the relief of a landscape. The term includes only the physical features, unlike topo-graphy which also includes the man-made features of a landscape. The study of relief features is called geomorphology.

relief map A map which shows relief features, by contours, or shading and colouring. Many Swiss maps use **hachuring**, which is a system of lines drawn down the slopes. The steeper the slope, the more numerous the lines. Some ordnance survey maps use hill shading, in which hills are shaded on one side to give a pictorial impression of relief. They are drawn as though a light were shining from the northern edge of the map, and so all south-facing slopes are shaded.

relief rainfall see **orographic rainfall**

remote sensing The use of spacecraft and satellites to take photographs from great heights. Not only conventional photographic equipment is used, but also various infrared and electronic devices which produce images that may need specialist interpretation. A common use of satellite photographs is in weather forecasting, because cloud height and thicknesses can be seen, as well as temperatures in the oceans. Remote sensing enables information to be gathered from inaccessible areas and the photographs can be used to show the changing vegetation patterns from season to season, or from year to year.

rendzina A soil type which occurs on chalk and limestones. The A horizon is often loamy though quite thin, and the B horizon contains fragments of chalk or limestone rock. Rendzinas are found in the chalkland downs and wolds, as well as in parts of the Cotswolds and Pennines, where grassland vegetation has covered the land for a long time. Many rendzinas are changing because they are being ploughed and used for cereal production. Rendzinas are found in many other countries, too, including Yugoslavia, western Ireland, parts of eastern Texas and Alabama.

renewable fuel Any form of energy of which there is an infinite supply. Hydroelectricity, wave power, wind power and tidal power are all renewable forms of energy, supplies of which will be available to future generations. Nuclear power is non-renewable at present though it is possible that scientists will develop a renewable method of creating energy from nuclear fuels.

renewable resource A resource which can be replenished, such as wind and water. Soil is a renewable resource if it is carefully managed. Careless or thoughtless farming may destroy the soil and cause soil erosion, and in this case the soil cannot be regarded as a renewable resource. By replanting as many trees as are cut down, timber can be a renewable resource. Selective cutting enables forest to survive permanently, and in some places wholesale clearance is followed by immediate replanting.

representative fraction The scale of a map expressed as a fraction of the actual size of the area represented. An ordnance survey map of 1:50,000 means that one unit on the map represents 50,000 of the same units on the surface of the earth. The units are usually centimetres or inches. The smaller the number the larger the scale. The second most popular size of ordnance survey map is the 1:25,000, which shows far more information than the 1:50,000, though for a smaller area. Atlas maps have scales of varying sizes but may use 1:1,000,000 in some cases.

reservoir A lake used for storing water. Sometimes it may be a natural lake, but generally reservoirs are man-made lakes which have formed behind dams. They often provide hydro-electric power, as well as water for human or industrial consumption, and possibly for irrigation too. Reservoirs can be found in Britain; for example, Hawes Water in the Lake District, Lake Vyrnwy in central Wales, the Elan reservoirs, and many others. Large reservoirs are found in many countries, often where there is a marked seasonal distribution in rainfall. The heavy rains of the wet season are stored for use in the dry season; for example, in Lake Nasser on the Nile, the Damodar valley reservoirs in India, or the reservoirs in the Snowy Mountains in Australia.

reservoir rock A porous rock in which the pores have been filled by water, natural gas or oil. If the rock contains water, it will be described as an aquifer. Reservoir rocks containing oil or gas need a trap to prevent the mineral escaping; this often occurs in anticlinal formations. The oil or gas moves to the top of the anticline, but cannot escape because there are impervious layers above and below. Traps also are found near salt domes and fault lines. Reservoir rocks are sought by geologists when they are looking for oil or gas deposits.

residual deposit Weathered and eroded fragments which are left behind. Weathering occurs in situ and leaves fragments of rock on hillsides as scree, or on top of limestone or chalk as terra rossa or clay with flints. Erosion may involve some movement, and residual deposits may be dumped by ice or rivers.

residual hill A small but steep, isolated hill; the last remnant of a larger mass of high ground. It may be the middle of a compressed anticline, a metamorphosed area, or the puy in a volcano.

resistant rock A relatively hard rock which can withstand weathering and erosion more successfully than neighbouring rocks. Resistant rocks often form higher ground, as the softer neighbouring rocks are worn away. They may be hard because of their mineral content, or possibly because the particles are well cemented, or because of the absence of joints and lines of weakness. Whether or not a rock is hard and resistant will depend on the adjacent rocks; resistance is relative. New red sandstones which are adjacent to old red sandstones in southwest England seem to be soft, but when found alongside the clays of the Midlands they are resistant, and form higher ground than the softer clays.

retailing The selling of goods. Retailing is a small scale activity, in comparison with wholesaling, and is generally carried on from a shop. The term retailing is sometimes extended to more than merely sales from shops, and can include such activities as hairdressing, banking, catering and the world of entertainment. Most retailing is found in city centres (the central business district). Larger settlements have a greater number of retail functions than can be found in a small settlement. Small villages may only have one or two retail outlets, a small town may have 50 to 100 and a large town may have several hundred. The different types of central places outlined by Christaller can be quantified by a study of retailing outlets.

revolution A complete circuit of the earth around the sun. Each revolution takes 365.25 days. Because of the tilt of the earth's axis at 23.5°, the revolution causes the four seasons experienced in temperate latitudes. In spite of the tilt and the revolution, there are no major seasonal changes near the equator.

rhine Also **rhyne** A drainage ditch found in wet fenland and marshy areas, such as the Somerset Levels.

rhyolite see **volcanic rock**

ria A drowned river valley in a hilly landscape. Where a river valley has been flooded, either by a rise in sea-level or because the land has sunk, quite small rivers can develop large estuaries. The cross-section of a ria is V-shaped, in contrast to the cross-section of a fjord which is U-shaped. Rias often have branches, which are old tributary valleys, and the estuary becomes wider and deeper towards the sea. They provide sheltered harbours, and have been used for fishing in the past, but are important for sailing in many places now. Devon and Cornwall have many examples. On a larger scale, when a discordant coast has been flooded, large inlets are created and these are also called rias. Good examples can be seen in southwest Ireland and northwest Spain.

ribbon development Urban expansion along the roads leading from large towns. Linear expansion of this kind was especially common in Britain during the 1930s. At that time the land between two main roads was usually left as rural greenery, but it was widely utilized for building housing estates during the 1950s and 1960s.

ribbon lake see **finger lake**

rice A cereal, *Oryza sativa*, which is most commonly found in monsoon regions, but also grows well in equatorial climates. Rice is grown in the United States along the Gulf coastlands, and also in the Po Valley in Italy, the Camargue in southern France and in places along the Mediterranean in Spain. It gives a high yield per acre, and two or three crops can be grown per year if heat and water are available. The most productive areas are on the deltas of Asiatic rivers such as the Si in southern China, and the Ganges, Irrawaddy, Mekong, Menam and Red rivers. Rice feeds more people than any other cereal. In the last twenty years productivity has been increased by the development of new varieties by the International Rice Research Institute in the Philippines. The miracle rice was I.R.R.I. 8, and many new varieties have been grown since. Each new variety encounters difficulties with pests after a few years, and so it is a permanent research task to produce different varieties.

Richter scale A logarithmic scale for measuring the magnitude of an earthquake. It was developed by American seismologist C.F. Richter during the 1930s, and has now replaced the Rossi-Forel and Mercalli scales. Earthquakes are measured on a scale of 0–9, and a measurement of 8 or over represents an earthquake of great intensity.

ridge A narrow elongated area of high ground with low ground on both sides. The sides are generally quite steep, and a ridge is likely to consist of harder rock than the neighbouring rocks.

ridge of high pressure An elongated area of high pressure which is an extension from an anticylcone. It is similar to, though rather wider than, a wedge of high pressure. It gives weather similar to that of the anticyclone but unlikely to last as long. Britain is often affected by northward pointing ridges which extend northwards from the Azores high. They fit between depressions, and bring a short dry and sometimes sunny spell, especially to southern England. In winter ridges of high pressure may spread out from Europe towards Britain; they have most impact on the eastern parts of the country, and bring very cold weather.

rift valley A valley resulting from earth movements along fault lines, where two or more roughly parallel faults have caused land to sink by downfaulting. Normally greater in length than width, they can be quite small or very extensive; for example, the Great Rift Valley of east Africa extends for nearly 5000 km, from Syria along the Jordan Valley, the Dead Sea, the Gulf of Aqaba, the Red Sea, and south to lakes Malawi and Tanganyika. The part of Africa east of this line is splitting away from the rest, and will gradually drift east to form a new land mass. Other rift valleys include the Rhine valley near Strasbourg, the Clwyd valley in north Wales, the Central Valley of Scotland and the Great Glen between the Grampians and the North West Highlands of Scotland. The steep or near vertical slopes caused by the faulting are quickly worn down by erosion; even so, rift valleys are often very steep sided. Also called **graben**.

rill A small channel formed by water erosion.

rime A type of frost caused by supercooled droplets which freeze on to exposed objects. Rime drifts along on gentle breezes and freezes on to trees, telegraph poles, or any solid substances. It occurs only on the windward side of objects, building up feathery shaped crystals.

ring city A type of conurbation consisting of a circular urban development enclosing a central green belt. The best example is **Randstad** in the Netherlands.

ring road A road which goes round a city in order to divert traffic away from the city centre and reduce congestion. Many cities have ring roads; for example, Birmingham and Norwich. London has the M25, in addition to the North Circular and South Circular routes.

river A channelled flow of water, which moves downhill because of gravity. Most rivers flow into the sea, although there are some which flow into lakes, and some flow into deserts, where they evaporate to leave salt flats or very saline lakes. Those which do not flow to the sea are called areas of inland drainage. Most rivers flow throughout the year, although the volume of water will vary. Some rivers, especially in deserts and regions with marked variations in rainfall, may dry up for a few months each year. The source of a river is generally in an area of high ground, and it may be a spring, a lake, a marshy patch, or a place where the water-table reaches the surface. A river's route is called its course. From its source it generally flows through mountains then through an upland area and finally onto a lowland, before entering the sea in an estuary or through a delta. The upper parts of a river are usually rocky and turbulent, and the narrow valley is V-shaped with steep sides. There may be waterfalls to interrupt the long profile. In the middle course the valley becomes wider, and meanders begin to develop. Bluffs and small flood plains may be forming. The lower course passes through a wide flood plain; the meanders are larger, and some may have been abandoned to form oxbow lakes. The longest rivers in the world are Nile, 6649 km; Amazon, 6276 km; Mississippi-Missouri, 6111 km; Ob, 5570 km; Yangtse, 5520 km; Hwang, 4672 km; and Zaïre, 4667 km.

river basin see **catchment area** (def. 1)

river capture The capture of a river's water supply by a second river, which works back uphill by headward erosion and intercepts the flow. Also called **river piracy**.

river cliff see **bluff**

river profile A drawing of a section of a river, whether a long profile or a **transverse profile**, which is cross-sectional and can be very variable. In the hilly upper section of the river, the valley is likely to be steep, narrow and V-shaped. Further downstream, as the river grows, meandering begins, the valley widens and the profile becomes more U-shaped. Once out on to the plain, the cross-section is very flattened and U-shaped. In all parts of the valley the cross-section is likely to be asymmetrical.

river regime The pattern of seasonal variations in the volume of a river.

A river regime can be shown in the form of a graph. In Britain, rivers generally show a greater flow in winter, when there is more rainfall and less evaporation, although occasionally in late summer there will be high points caused by torrential rain during thunderstorms. In monsoon and savanna areas the high points on the graph will be in summer, which is the rainy season, and some rivers may dry up altogether in the winter. The regime of a river is determined largely by the rainfall, but the local rock type will also have some influence. In granite areas there is always a rapid surface run-off, whereas in areas of chalk, limestone, or porous sandstone much of the water goes underground and is released gradually during the next few days. The variations in speed of through flow obviously influence the quantity of water reaching the river. When urban areas are built with a network of drains, it often means that rainwater gets into the river far more quickly than previously, when there were trees, soil and rocks to slow down the rate of run-off.

river terrace An old and abandoned flood plain. When rejuvenation gives a river renewed powers of erosion, it is able to cut its valley down to a lower level. When this happens, the former flood plain is left at a higher level, and a new plain is created. The old plain is called a terrace, and in many valleys flattish areas can be seen above the level of the river. They are usually quite small. Terraces are numerous along the Thames, and parts of central London are located on terraces. Terraces often occur in pairs, with one on each side of the river. In Britain there are often two or three terraces at different heights, probably resulting from changes of sea-level during the glacial and interglacial phases of the ice age.

roaring forties The area between latitudes 40°S and 50°S where the westerly winds blow uninterrupted by any land mass. The winds are persistent and strong, much more so than the equivalent winds in the Northern Hemisphere, which blow over Britain and western Europe, and on to northwest United States and British Columbia in Canada. In the roaring forties a series of depressions continually moves from west to east, bringing windy and wet conditions but quite mild temperatures.

robber economy The utilization of a resource which cannot be replaced.

This is inevitable in the case of minerals, but it has also been practised in other situations. For example, soil was used wastefully by early settlers in the western United States. Forestry can be a robber economy, unless trees are planted to replace those which have been cut down, and frequently fishing has been another example, but there are now international agreements to restrict the size of catches.

roche moutonnée (*French*) A rock which has been scraped smooth by ice on one side, but has been plucked into a jagged shape on the other, downhill side. Roches moutonnées can be only a few metres in height, but some are much bigger, reaching 30 to 40 m. They are found in glacial valleys; when viewed from up the valley they will appear smooth and gentle, but viewed from downstream the jagged rocky edge will be visible, and the landscape will look very different. The smooth surface will have scratch marks, or striations, caused by ice and small rocks scraping over it. The jagged downstream side will have suffered from freeze-thaw activity. It is said that roches moutonnées were so named because they look rather like sheep lying down, or perhaps like old fashioned wigs called moutonnées. A good imagination is required to see the resemblance!

rock A collection of minerals which have been cemented together. Rocks make up the surface of the earth, and they are very varied in appearance, hardness and mineral composition. All rocks can be classified into one of three groups: sedimentary, igneous, or metamorphic. Sedimentaries consist of sediment, generally carried by rivers down to the sea, where it all accumulates, is compressed and uplifted to form new rocks. They are composed of land sediment together with fragments of dead sea creatures. Igneous rocks are those which result from volcanic activity, either underground (intrusive), or out on the surface (extrusive). Metamorphic rocks were either igneous or sedimentary, but they have been changed by the effects of heat or pressure. The heat comes from igneous activity, and the pressure is the result of folding. The most intensive metamorphism occurs when heat and pressure are experienced in the same locations.

rock flour Fine particles produced by abrasion beneath a glacier or icesheet. As the ice moves downstream, the rocks it is carrying scrape and grind the bedrock, and tiny fragments of rock flour are produced. Much of the rock flour is transported by sub-glacial streams and contributes to the load of glacial rivers, which are often inky in colour as a result.

roll-on-roll-off A ferry service which takes vehicles. Cars, lorries and coaches can drive directly into the ship at one end, and off at the other on completion of the crossing. Examples include Newhaven to Dieppe, and Dover to Calais.

root crop A plant which is grown for its edible roots. Potatoes are particularly important for human consumption, and sugar beet is an important source of sugar. In tropical lands manioc, yams and sweet potatoes are vital to many groups of people. Turnips, swedes and several other root crops are a common source of cattle feed.

ropy lava An extrusion of lava which has solidified in extended linear fashion, like pieces of thick rope. A thin skin forms on the lava while it is still moving, but the molten liquid beneath the surface continues to move for a time to produce an elongated shape. Also called **pahoehoe**.

Rossi-Forel scale A scale for measuring the intensity of earthquakes devised in 1878 by M.S. de Rossi and F.A. Forel. It was commonly used until 1931, when the modified Mercalli scale became more popular. Both have now been superseded by the Richter scale.

Rostow's model A model describing the stages of economic development which all countries will pass through. It was produced in 1955 by W. Rostow, who believes that all countries will follow a similar pattern of development as they progress from a traditional agricultural economy. There are five stages to his model:

Stage 1. The traditional society is still largely agricultural, often at a subsistence level, and there is little utilization of resources. Britain was at this stage several hundred years ago, but some areas of the world are still at this level.

Stage 2. Preconditions for take-off. Some modern methods of agriculture may be introduced, and outside influences such as colonialism will bring about changes. There will also be the beginnings of exploitation of resources.

Stage 3. Take-off. This means the beginning of industrial development, which may be initiated by the use of resources, such as coal in Britain and western Europe in the early 19th century, or the exploitation of oil in the Middle East in the 1960s. Earnings increase, and there is money available for investment. Agriculture becomes more commercialized.

Stage 4. The wealth created in stage 3 provides the basis for expansion of industry. This is called the drive to maturity, and growth occurs in all sectors of the economy. Money is available to buy luxuries and consumer goods, which provides a boost to many other industries. Technology and technical skills increase.

Stage 5. The stage of high mass consumption. Urbanization has resulted in well over 50% of the total population living in towns, and this figure reaches as much as 90% in some countries. The affluent society, as in the United States, western Europe and Australasia, spends more on consumer goods and services. The major area of employment is in the tertiary sector, with an increasing number of quaternary jobs too.

The older developed countries have passed through the stages at a fairly slow rate, but many of the Third World countries are progressing through the five stages in a few decades. There may be different regions of the same country in each of the five stages at the same time; for example, Brazil, Mexico, India.

rotation of crops A system of farming in which each field is used for a different crop or animal every year. There may be a planned cycle of planting over a period of two to eight years. The first well-known method of crop rotation was developed in Norfolk. The four-year cycle was a root crop – generally turnips – followed by barley, then grass and clover, and finally wheat. The rotation gave two cereals and a good fodder crop, as well as grass which could be used as pasture, and clover which would restore nitrogen to the soil. It was a great improvement on previous systems, which had involved leaving the land fallow to allow it to rest and recover. It allowed all land to be used in a productive fashion, while maintaining the quality of the soil. The main disadvantages of crop rotation are that farmers are involved in a large range of activities and therefore cannot afford specialized equipment, and that as they have many small fields they cannot use large-scale methods. On the other hand, the advantages are that the quality of the soil is maintained and that animal manures are available, which will reduce the dependence on chemical fertilizers. Weeds, pests and diseases are more easily kept under control, again reducing the need for chemical controls.

rotation of the earth The turning of the earth about its own axis. One complete rotation takes 24 hours, during which day and night are experienced, as different parts of the earth turn to face the sun. Because of the tilt of the earth's axis at 23½° the length of day and night varies. Without the tilt, they would be of equal length. At the equator day and night are approximately of equal length throughout the year. At the North Pole there is 4½ months' daylight and 4½ months' darkness, and the rest of the year is twilight. The same is true of the South Pole, although the periods of day and night are reversed. All places north of 66½°N and south of 66½°S receive at least one 24-hour period of continuous daylight. The length of day in Britain varies from summer to winter. However, in a complete year, all parts of the earth receive exactly the same amount of daylight, although the daily and monthly distribution is variable. On the equinoxes – March 21st and September 23rd – everywhere has 12 hours' darkness and 12 hours' light.

rough grazing Low quality and unimproved grassland. Some moorland, scrubland, marshland and mountain pasture is described as rough grazing. It is often unfenced and used for extensive grazing of sheep, or occasionally beef cattle. Rough grazing can be improved by the addition of fertilizers and better, sown grasses. Some parts of the world do not distinguish between different qualities of pasture, and merely divide farmland into arable and pastoral.

rubber An elastic material, obtained from the latex of various species of *Hevea* and *Ficus*. The most important rubber tree is *Hevea brasiliensis*, which originated in the Amazon basin, but is now grown on a larger scale in plantations in Malaysia, Indonesia, Sri Lanka and elsewhere. The native rubber trees are still used in Brazil, but production is very small, and no plantations have been developed. Manaus in the middle of the Amazon basin remains a small market, but a century ago it was the rubber centre of the world. Synthetic rubber can now be produced in chemical works, using petroleum as the raw material. The United States and West Germany manufacture large quantities of synthetic rubber, and so the demand for natural rubber has declined.

rudaceous rock A coarse-grained sedimentary rock. Rocks which are derived from land sediment are split into three main groups according to grain size. If the particles are coarse, like large grains of sand or gravel, the resulting rock when consolidated is rudaceous. Examples include breccia, conglomerate, gravel, scree and some types of stony boulder clay. Less course-grained rocks are classed as **arenaceous**, which means sandy, and fine-grained sedimentary rocks are known as **argillaceous** rocks, or clays.

run-off see **delayed run-off, surface run-off**

rural depopulation The movement of people from rural areas into the towns. Such movement has been going on for centuries, but has quickened in more recent times. People are attracted to the towns by the bright lights, more jobs, better pay. The movement grew rapidly in Britain after the Industrial Revolution, and urban areas expanded dramatically. As more jobs and more money became available, rural inhabitants left the land, so that many farms changed to more pastoral activities, which required less labour, but were often equally profitable to the landowners. As the industrial developments grew, machines were provided to lighten the burden of labour on the farms, and this enabled more people to leave and drift in to the cities. Other European countries saw similar movements taking place. In more recent times, especially since 1950, there have been large scale movements from rural areas into the cities in Third World countries. Unfortunately there are not enough jobs available, and so many of the migrants are very poor and have to live in shanty towns.

rural settlement A settlement in a rural area, typically a small village or hamlet. Formerly, rural settlements were associated with working in agriculture, but in western Europe and North America this is not always true, as only a very small proportion of the working population is employed on farms. In the United Kingdom the figure is now below 2%. Many rural settlements are now occupied by city dwellers, as around Greater London. The distribution of rural settlements in modern and industrialized environments cannot really be related to the Christaller model.

rural urban fringe   The transitional area between town and country. It is not always clear where an urban area ends, and even beyond its limits it will still be very influential on the people living in the surrounding rural areas. The idea of a rural-urban continuum is found in European countries, North America and Japan. There is now a graduation from urban to rural areas, as urban influences stretch out into the country, and many rural dwellers have their own transport, which enables them to use the facilities of the town. The effect of distance has decreased because of the increase of mobility. Rural areas depend on towns and cities for jobs, supplies of food, shops, hospitals, schools and also for administration, which is based in the larger towns. In Third World countries the situation is still similar to Europe of a century ago. Many people are still involved in agriculture, and many do not have transport and mobility.

S

saddle A slightly lower area between two summits. It is often flattish, and provides a gap for a route between two valleys. It is similar to a col, but probably slightly broader.

saeter Also (*Norwegian*) **seter** An area of mountain pasture. Located above the forests of the mountain side, it will be used for summer grazing. Most families from the villages in the valleys have a hut on a saeter, and one or two members of the family move up to the hut to look after the cattle while they are on the pasture. The cattle are milked, and butter or cheese is made, although some milk may be transported down to the valley. This use of mountain pasture is very similar to alpine farming in Switzerland, France and Austria.

sage brush A type of thorny xerophytic scrub vegetation, which can survive in semi-arid locations. The southwest United States has large areas of sage brush, and Nevada is known as the 'sage brush state.'

Sahel 1 A large area of Africa to the south of the Sahara. It stretches from Senegal and Mauretania in the west to Sudan and Ethiopia in the east, and includes parts of Burkina Faso, Nigeria, and many other countries. It has been noted for famines in recent years, and the subsistence farmers living in this area have found life very difficult. Overpopulation, as well as too many animals, has prevented the resources of the area from supporting all the people. Overgrazing has ruined the vegetation, and the land has been exposed to soil erosion. The bare land has lost its soil, and so no vegetation can grow. The Sahara desert has been expanding southwards into the new man-made desert. Desertification has been accelerated by the fact that most of the local people use firewood for fuel, and if any trees can be found they are used as fuel, so that there is nothing left in the ground to protect the soil. Heavy rain will wash away the bare soil, and wind can blow away the fine particles. 2 The semi-arid lands in Tunisia, along the margin of the Sahara. Although not a wealthy area, this region does not have the same kind of problems as the land to the south of the Sahara.

saki A means of lifting water from one level to another. It is rather like a seesaw with a bucket at one end, and is an important method of irrigation in parts of the Nile Valley and the Middle East. Also called **sakiyeh**.

salina (*Spanish*) A saltpan. Salinas often form in desert or semi-arid regions where a stream has flowed into an area of inland drainage, and then evaporated, leaving tiny deposits of salt behind. The deposits can accumulate for hundreds of years to form large salty areas, such as are found in the interior of Australia, in parts of the southwest United States and on the Andean plateaux of Bolivia.

salinity The degree of saltiness, as of the oceans. The average proportion of salt in the sea is about 35 parts per 1000, or 3.5%. In the Red Sea, because of high evaporation rates, the figure is over 40, but in the Arctic it falls to less than 30. In the northern parts of the Baltic it is less than 10, and in the Dead Sea the figure is about 250. The salts which are found in the oceans include sodium chloride (common salt), which accounts for about 75%, magnesium chloride, magnesium sulphate and calcium sulphate.

salt Sodium chloride, $NaCl$. It is found in sheets and layers around salt lakes, as well as in old strata, where it is a relic of desert conditions of millions of years ago. The largest deposits in Britain are in the Permo-Triassic rocks of the New Red Sandstone in Cheshire, which were formed nearly 250 million years ago when Britain was similar to the Sahara. Salt often folds up to form domes, which are not only important sources of salt, as in Texas and neighbouring states, but also make good reservoirs for oil and natural gas.

saltation A process whereby sand grains moved by the wind bump into other grains, which are then moved forward as well. In this manner sand is moved downwind in desert areas, as well as in areas of dunes. The same term is used for the movement of sand along the bed of a river, in which the flowing current dislodges grains of sand and pebbles.

salt lake A saline lake located in a desert area, probably in a region of inland drainage. As the water evaporates the salinity is likely to increase, and around the shores of the lake, layers of salt will solidify. Examples can be seen in all desert areas but amongst the largest are the Great Salt Lake of Utah, in the United States, and Lake Eyre in Australia. If there is a spell of exceptionally wet weather, much of the salt may be dissolved and washed away.

salt marsh An area of coastal marsh which is flooded by the sea from time to time. There are many locations in Britain where salt marshes occur, such as Chichester harbour, in Essex. Salt marshes are also found around salt lakes in deserts, and near large lakes and estuaries in warm regions, such as the Camargue in southern France or the Guadalquivir delta in southern Spain. Special salt-resistant plants colonize the areas; they trap silt and help to build up the land. Many of the coastal marshes can be reclaimed once the land has become well vegetated.

sand-dune A ridge of sand formed in deserts and along coastlines wherever there is loose material which is not anchored by vegetation. Sand particles are transported by the wind and gradually move in a downwind direction. Sand-dunes form along sandy shorelines, and dry sand is gradually blown inland. It piles up into a series of ridges, with lower hollows or slacks between. Marram grass often colonizes the dunes and helps to anchor them. In Britain there are many coastlines with sand-dunes; for example, at Borth in Central Wales, and near Southport. In deserts there are often large expanses of sand sheet, but there are two distinctive dune shapes; these are seifs and barchans.

sandstone A sedimentary rock, in which the grains of sand are 0.02 to 2 mm in diameter. If the grains are larger, the rock is a rudaceous rock or gravel, and if smaller the rock is silt or clay. Sandstones are arenaceous rocks, and are very variable in character as a result of differences in grain size, chemical content and cementing materials. Most sand grains contain quartz, which is hard and resistant. Other minerals in sandstones include felspars and mica, and iron also occurs, which tends to give sandstones a reddish or brownish colour. Most sandstones are porous and produce poor soils for agriculture, though they can be improved quite easily with the addition of fertilizer. They give light soils suitable for vegetables, and other crops. Some sandstones are very hard and give rise to high ground; the millstone grit of the Pennines and the sandstone of Exmoor, are examples. Other sandstones, such as the New Red Sandstones of Cheshire, are soft and easily eroded, forming lowland areas. The majority of sandstones are in between the two extremes, and form undulating countryside, such as the hills of the Weald of Surrey, Sussex and Kent. Here there are several different sandstones, notably the greensand, which produces low hills rising above the clay valleys, but not rising as high as the chalk downlands. Most sandstones are formed by the accumulation of river sediment on the seabed. They are then compressed and uplifted to form new land. There are also a few sandstones which have been formed of wind blown materials, especially in desert regions.

sandur see **outwash plain**

saprophyte A plant which lives on decaying vegetable matter. Fungi are good examples of this type of plant. Saprophytes are vital decomposers in ecosystems.

satellite photographs There are numerous satellites circling the earth at present. Some are geostationary, which means they remain over the same point of the earth at a height of 36,000 km and travel at the same speed as the rotation of the earth. Others circle round the earth in a polar direction, and with each orbit pass over the equator about 30° west of the previous orbit; this is because the earth continues to rotate below the satellites. The photographs have daily use for weather forecasting, but can also be used in many other ways. For example, they may provide information for mapping land use, plotting water temperatures in the oceans or the position of icebergs. Satellite photographs are images recorded electronically rather than traditional photographs. Landsat satellites send down informaton on land use, and Meteostat gives information on cloud cover.

saturated adiabatic lapse rate The rate at which saturated air cools as it ascends. The average rate is 0.5°C per 100 m, but it shows considerable variation, depending on the air temperature and the amount of water vapour present. The rate is much lower than in dry air, because the moisture gives off latent heat when it condenses. See also **dry adiabatic lapse rate**.

savanna   An area of tropical grassland, found between the major desert regions and the tropical forests. Savannas occur mainly between 5° and 20° north and south of the equator, although in east Africa, savanna regions are situated on the equator as well. This is because the land is elevated, and the East African plateau is at the appropriate height for grassland vegetation to grow. Both the rainfall and temperatures are too low for tropical forest. The savanna vegetation is supported by the savanna type of climate. This is hot and wet in summer, when the sun is overhead; temperatures average about 25°C and the rainfall is 250 to 1250 mm. In winter the tropical high pressure system controls the climate, and so the conditions are warm but very dry. Temperatures are 15° to 25°C, and rainfall totals are less than 250 mm. In such a climate grass grows very well in summer – up to 2 m in height but it all shrivels in winter. Trees do not grow well, although a few acacias and eucalyptus are able to survive the long dry winters, and baobabs and bottle trees survive by storing water in their trunks. Trees are most numerous at the edge of the savanna which is nearer to the tropical forest. In the Northern Hemisphere, this is the southern edge; for example, in Nigeria and Venezuela. It is the northern edge for the Southern Hemisphere savannas, such as the Brazilian plateau and parts of Australia. Pastoral farming is the main activity on the grasslands, and this may be large-scale commercialized farming, as in Northern Territory of Australia, or subsistence and semi-nomadic, as for the Masai or Fulani in Africa. Coffee is grown on plantations in Brazil, and tobacco in Zimbabwe, and maize is generally the main food-crop grown. There are many wild life parks in African savanna areas, where giraffe, lions, and other animals can be seen. The savannas are sometimes called sudan areas in eastern Africa, and are known as *campos* in Brazil and *llanos* in Venezuela.

scale   The size of a map expressed as a ratio of the actual size of the land. See also **representative fraction**.

scar   A short steep slope, which is generally bare rock. The term is used in northern England, and it is generally associated with areas of carboniferous limestone.

scarp   see **cuesta**

scatter graph* A graph which shows the relationship between two variables. A line of best fit can be drawn on the graph to show the average trend, and then the extreme examples, above and below the average will be clearly shown. One of many ways in which a scatter graph could be used would be to put the population of settlements along the horizontal axis, and their height above sea-level on the vertical, to determine if there were a relationship between height and size. Also called **scatter diagram**.

schist A metamorphic rock characterized by thin plates or flakes of minerals, overlapping rather like the tiles on a roof. Almost any type of rock will become a schist if subjected to sufficient metamorphism, and it will be named after its most common mineral; for example, mica schist. Many examples can be found in northwest Scotland.

science park An area, generally on the outskirts of a town, in which a new industrial estate has been created. Most of the work undertaken in the area will be scientific and high technology, often involving high-level research projects. A most important science park is located near Cambridge, but many other towns have examples, which are usually located near to a university.

scree Loose fragments of rock which have accumulated on a hillside. They are likely to slide down the hill when lubricated. The rock fragments may be very varied in size, and a small amount of sorting will take place as the particles move downhill, the larger particles tending to move to the bottom. Scree forms more frequently on certain rock types, such as limestone. Freeze-thaw activity and the effect of changes in temperature are the most important processes involved.

scrub A type of vegetation found in locations which do not have rainfall throughout the seasons. Low shrubs and small trees up to 2 m in height are common, and the growth may be quite dense. Good examples are found in a variety of locations. There is the maquis in the Mediterranean lands, the mallee in Australia, and the chaparral in California. There is also much scrub along the margins of deserts, between the savanna and the more arid lands; for example in west Africa. There is much scrub in Brazil and also in neighbouring Paraguay, where the Chaco region is a good example.

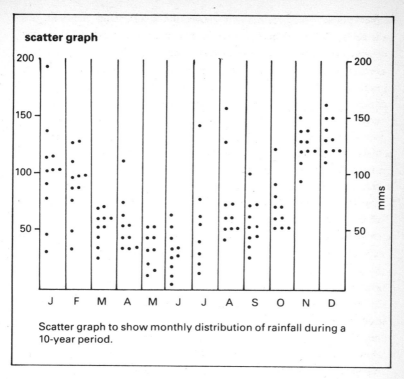

scatter graph

Scatter graph to show monthly distribution of rainfall during a
10-year period.

sea breeze A wind which blows from the sea on to the land. In tropical
areas it is usually a cooling breeze, as the sea is cooler than the land. In
temperate latitudes a wind from the sea can feel cool and raw, and can
be bitterly cold if coming from the Arctic or any icy sea. A daily sea
breeze often develops in settled anticyclonic weather. As the land heats
up during the day, air begins to rise. The land becomes warmer than
the sea, and, as the air pressure over the land is slightly lower, a wind
will begin to blow from the sea. See also **land breeze**.

sea-floor spreading see **ocean-floor spreading**

seasons One of the four periods of the year: spring, summer, autumn
and winter, which are characterized by different climatic conditions.
According to the Meteorological Office, winter is December, January
and February; spring is March, April and May; summer is June, July

and August; and autumn is September, October and November. In arctic regions, spring and autumn are very short, and summers are fairly cool. Near the equator there are virtually no seasonal differences. In the Southern Hemisphere, winter is during June and July, and summer is December and January.

sea surge A large movement of oceanic water, which is likely to cause large waves or high tides when it reaches land. Britain is affected ocasionally by a North Sea surge, when a large volume of water is moved southwards through the North Sea. If this coincides with high tide, it can cause considerable flooding. This is most serious in areas such as the Fens, south Essex, the Thames estuary and in the Netherlands, where there is much low-lying land. The serious flooding in 1953 was caused by a North Sea surge coinciding with a high spring tide, and large areas of land were flooded. Some of the water went up the Thames estuary, and nearly caused flooding in central London. Fear of this happening in the future led to the construction of the Thames barrier. North Sea surges are caused by deep depressions; the low pressure can have slight effects on water-level, and the strong northerly winds at the back of the depression blow extra quantities of water into the southern parts of the North Sea.

Secondary A geological era, now generally known as the **Mesozoic**.

secondary depression A small area of low pressure which is found on the edge of a major depression. It may just be a bulge in the shape of the isobars, or it may be a separate system with its own set of roughly circular isobars. Secondary depressions often become deeper and more intense than the primary depression, and in western Europe they often move around the primary in an anti-clockwise direction. Like all depressions, they gradually fill in and become weaker.

secondary industry An industry which processes materials produced by the primary sector. Secondary industry is considered to include the building industry.

secondary sector The part of the economy which deals with manufacturing and processing; for example, making steel, cars, or clothing. Employment in secondary activities increases as countries become industrialized, but decreases later as the tertiary sector develops. The tertiary sector provides services; the primary produces raw materials; and the quaternary provides information.

sector model One of the three classical models of urban morphology, described by Hoyt, in 1939. He used sectors as the basis of the divisions of land use in a city, and believed that transport routes would be a major influence on the growth of urban areas. He described his model of urban structure after analyzing land use patterns in 142 American cities. His five main types of land use were: 1 central business district, 2 industrial zones, 3 low-quality housing, 4 medium-quality housing, and 5 high-quality housing. He added his ideas to those put forward by Burgess in his concentric model, but it is still an oversimplification of reality.

sedentary agriculture Settled agriculture in a fixed location, as opposed to nomadic or shifting agriculture. All commercial crop growing has to be sedentary. With the exception of some transhumance movements, pastoral farming is also sedentary, unless it is for subsistence only.

sediment Rock fragments which have been formed by weathering and erosion. They can be very varied in size. Once broken down, they are moved by water, and in certain places by ice or wind, and will be transported largely into low-lying areas. Most sediment is eventually deposited in the sea.

sedimentary rock Any rock formed by deposition of sediment derived from pre-existing rocks, which may have been sedimentary, igneous, or metamorphic. Most sediment accumulates on the bed of the sea, having been dumped there by rivers, or accumulated there as dead sea-creatures fell to the ocean floor. The accumulated sediment will be consolidated and compressed. Earth movements uplift the sediments, and they may be tilted, folded, or faulted. The resulting rocks are sedimentary, and their type depends on their composition. The layers in sedimentary rocks are called strata; they may be a few centimetres or many metres in thickness. Sedimentaries consisting of land sediment are mechanically formed or clastic rocks, and are gravels, sands, silts, or clays, according to the size of the particles. There are also limestones (which consist of fragments of dead sea-creatures, sometimes mixed with land sediment), evaporites, which are saline deposits, and coal, which is accumulated vegetation, as well as coralline, which contains large quantities of coral, and chalk, which is a pure form of limestone, with very little land sediment.

seeding clouds The practice of putting more nuclei into clouds in order that more water droplets will condense on to the nuclei and form sufficiently heavy droplets to fall as rain. Dry ice and silver iodide are used as nuclei, and experiments have been conducted in the United States, Australia, southern Africa and the Middle East with the aim of producing rain in areas suffering from a shortage. In some cases, rain has fallen after cloud seeding, but it is not certain whether it is as a consequence, or just coincidence.

seif An elongated type of sand-dune, found in parts of Libya and other areas of sand desert. Often occurring in groups, seifs may extend for several kilometres. The narrow ridge will run parallel to the prevailing wind direction; when cross winds blow, the height and width of the seif can be increased.

seismic focus see **earthquake**

seismograph An instrument which records earthquakes. Some seismographs are so sensistive they can pick up seismic activity thousands of miles away. The vibrations are recorded by a nib on a revolving drum.

seismology The study of earthquakes.

self-sufficiency 1 Subsistence farming, in which smallholders or crofters grow enough food for their own requirements, and also make their own clothes. 2 The ability, especially of a developing country, to meet its own requirements for industrial goods.

selvas (*Portuguese*) A dense tropical forest found in the Amazon basin. There are similar forests in parts of central Africa and in southeast Asia. They are characterized by a rich variety of plants and insects, many of which are not found elsewhere. Large areas of the forests are being cut down in a wasteful and extravagant way at present.

séracs A very irregular surface of ice, generally found at the foot of an icefall on a glacier. It is the result of a crevassed section of a glacier becoming compressed when the gradient decreases, or the speed of ice movement is reduced. Pillars and pinnacles of ice stand up several metres above the general level of the ice, and make an almost impenetrable barrier.

sericulture   The rearing of silkworms for the production of silk. It is necessary to grow mulberry trees, as the silkworm lives on mulberry leaves.

serra   (*Portuguese*) A range of mountains.

service   Any of the tertiary industries, which help or serve the general public, including buses and trains, shops, banking and insurance. Large towns have a larger number of services than can be found in smaller settlements.

settlement   Any collection of dwellings, ranging from a small hamlet with two or three families to cities the size of London or New York. In rural areas there is often a small settlement every few miles, but larger urban areas are more unevenly distributed. There was generally a reason for the original choice of site, even though the reason may no longer apply, or the settlement may have grown too big for its original site. For example, water supply commonly dictates choice of site for a settlement, but if a large town develops the water supply will probably become insufficient. For this reason, most large towns, such as Birmingham, Manchester, Los Angeles, etc., have to obtain much of their water from many miles away.

settlement hierarchy   A system for classifying settlements according to their size and influence, as propounded by Christaller in his central place theory.

shade temperature   A temperature reading taken in the shade, usually in a Stevenson screen. Official temperature readings are always shade temperatures. They exclude the effects of sunshine and shade differences, and the effect of wind. The reason for this is to standardize methods, and to enable fair comparisons to be made. By using a whirling hygrometer, it is possible to overcome some of the effects of sunshine and wind, if temperature readings outside a Stevenson screen are required.

shading map   see **chloropleth map**

shaduf   A simple lifting device widely used for irrigation in the Middle East and elsewhere. It consists of a long suspended pole with a weight at one end and a bucket at the other. Manpower is used to dip the bucket into a well and lift it out again. Once out of the well it is swung round and emptied into a ditch or trough. It is similar to a saki, but is used for obtaining water from wells, rather than from rivers.

shaft mine A mine which uses a vertical shaft to reach and extract the mineral. This is likely to be more expensive than adit or open-cast mining.

shale A type of sedimentary rock which is similar to clay. It is fine grained and consists of thin layers or sheets. Each layer probably represents a period of deposition, after which there was a slight pause or change in sedimentation. The layers have been consolidated, but shales can be split easily into horizontal pieces.

shanty town An area of unplanned and spontaneous development in an urban area. Shanty towns are characteristic of large cities in the Third World, where thousands of migrants drift in from the rural areas in search of employment, which is generally not available. Many of the people have no money and nowhere to live. They build simple and often very primitive houses out of cardboard, beaten out and flattened petrol cans, or any other materials they can find. The shacks are often constructed at the edge of the built-up area, or on any spare piece of land which can be found in the city. For example, old quarries have been used in Rio de Janeiro. Derelict building sites, steep slopes thought to be unfit for proper housing, land near sewage outlets, and old rubbish dumps have all been used. In Calcutta there have even been squatters in a pile of large drainage pipes which were lying around. Shanties have high densities of population, which create problems of hygiene as well as crime. There are often no water supplies and no electricity. However, in many cases all over the world, shanties have been improved by the efforts of the people living in them. Stronger and more permanent housing is built by people who have managed to get a job and earn some money. In many cases a water supply is provided, probably just one pipe and one tap for one or two rows of houses. The gradual improvement of living conditions has meant that some shanties have been turned into real suburbs. Once the occupants are given legal status and official authority to reside on the land, they work very hard to raise their standards. Most Third World countries still have shanties, and possibly as many as one-third of the urban inhabitants of these countries are classified as shanty dwellers.

sharecropping A system of farming in which the landowner is given a percentage of the crop harvested instead of a rent in cash. The share will vary from place to place, but is often as much as 50%. In some cases the landlord, who is likely to be wealthy, may provide machinery and seeds for the sharecropper to use. Sharecropping was used in the southern United States when slavery was abolished, but is also found in many other countries, including France, where it is called **métayage**.

sheep track A very narrow terrace in an area of softish rocks, such as glacial drift, where solifluxion has been taking place. In some of the hilly parts of Britain they are used by sheep as walkways and can be worn to look like narrow footpaths. Although they are referred to as sheep tracks, their formation was quite unrelated to the sheep. Also called **terracette**.

sheet erosion A form of erosion in which the surface materials are all removed. It may be the result of wind blowing away all the fine particles, or rain wash carrying away the particles after sheet flooding. Sheet erosion occurs only if there is no vegetation cover. When it occurs, it removes the fine particles of soil, leaving the coarse less fertile parts behind. Once the land has been denuded by sheet erosion, it is likely that future rainstorms will start to cause gully erosion as well. Sheet erosion has been most serious in areas where all the vegetation has been removed by ploughing lands which were too dry, or by over-grazing. The dust bowl and parts of the Sahel have been particular sufferers from sheet erosion. Also called **sheet wash**.

shield An area of old hard rock, such as the Canadian Shield or the Baltic Shield. The Canadian Shield is located between Hudson Bay and the Great Lakes. The Baltic Shield is found in southern Finland, and in parts of Sweden and western Russia.

shield volcano A volcano which has been built up by eruptions of fluid basic lava. Because the lava has been free flowing, the slopes of the volcano are quite gentle, allowing the lava to flow out over a large area. Eventually, a massive mountain can be created. The largest example is Mauna Loa in Hawaii which has risen over 9000 m from the sea bed. It is higher than Mt Everest, but half of its height is below sea-level. Its base covers an area of nearly 50,000 sq km. The name shield is derived from the fact that basic volcanoes are shaped like an upturned saucer or a shield.

shifting cultivation   A style of farming found in tropical forest areas in South America, central Africa and parts of southeast Asia. The people live in a village community, and they make clearings in the surrounding forest, in which to grow crops such as manioc, yams, sweet potatoes, maize, and beans. Once an area of forest has been cleared – usually by burning – it will be used for growing crops for two or three years, by which time the soil will be losing its fertility. Then the clearing will be abandoned, and the forest can regrow, giving the soil a chance to recover. Meanwhile, a new clearing will be created elsewhere in the forest. The system works well if population densities are low, but if the land has to be reused too quickly – say in less than 30 or 40 years – the soil will not have recovered sufficiently for crops to grow successfully. Also called **bush fallow, chitamene** (in Zambia).

shingle   Pebbles which accumulate on a shore. The stones have been smoothed and rounded by water action and by attrition. Shingle sometimes builds up ridges, as at Chesil Beach near Weymouth.

ship canal   A canal which can be used by ocean-going vessels. Panama and Suez are outstanding examples. There are ship canals on the St Lawrence Seaway in North America, the Welland Canal near Niagara and the Soo Canals between lakes Huron and Superior. In Britain there is the Manchester Ship Canal, and in the Netherlands Amsterdam is linked to the North Sea by the North Sea Canal.

sial   The lighter, granitic layer of the earth's crust consisting largely of silica and aluminium. The layer of sial rocks forms the continents, which sit on top of the sima layer. The sial is less dense than the sima, and is generally not carried beneath the surface when subduction occurs. See also **sima**.

sierra   (*Spanish*) A range of mountains.

silage   A fodder crop which has been harvested whilst green. It is generally stored in a silo, and provides a very nutritious winter feed for animals. Drying and haymaking are eliminated by the use of green fodder, which is particularly helpful in wet climates such as in Britain or Scandinavia.

silicon An element which comprises about 28% of the earth's crust. It is the second commonest element in the crust, after oxygen, with which it combines to form silicon dioxide (SiO_2), generally known as silica or quartz. It is also found in felspars, micas and many other minerals.

sill 1 The lip or threshold of a corrie. It often acts as a dam, so that after the ice has melted a corrie lake is formed. 2 An igneous intrusion. It is a thin sheet which is almost horizontal and is concordant to the pre-existing structure, which means that it has squeezed in between the bedding planes, and then cooled and solidified. Sills can be a few centimetres, or many metres in thickness.

silt A sedimentary deposit laid down in a river or lake. It consists of particles which are finer than sand but coarser than clay, with a diameter ranging from 0.002 mm to 0.02 mm. Consolidated silt forms siltstone, one of many argillaceous rocks.

silviculture The cultivation of trees for commercial reasons; forestry.

sima The part of the earth's crust which makes up most of the ocean floors and also rests beneath the sial of the continents. It consists mainly of silica and magnesium, and is denser than the sial. See also **sial**.

simulation The use of a model to imitate reality. Humans can use and manipulate the parts of a model in a fairly subjective manner. Computers can be used in simulation, and they will be objective. Simulations can look into the future and make predictions about what is likely to happen. Students of the greenhouse effect often try to construct models of the atmosphere to simulate the present-day situation, and then run the models to see what is likely to happen to the world's climate in the future.

sinkhole A hollow or hole in an area of limestone where solution has enlarged a joint. There may have been some collapse of surface material. Water may seep down a sinkhole after periods of heavy rain, but there will not be a permanent flow of water down the hole. Some sinkholes are large enough for potholers to descend. Also called **sink**. See also **swallow-hole**.

sirocco A warm dry wind which blows northwards from the Sahara towards the Mediterranean. It generally blows ahead of a depression passing along the Mediterranean, and can last for one or two days. It is most frequent in spring when the depressions are most active, and the Sahara has already become very hot. The sirocco often has a low relative humidity; it can cause plants to shrivel, and is unpleasant for humans. Sometimes it carries large quantities of dust from the Sahara. If it crosses the Mediterranean, it will pick up moisture and bring hot, humid, and rather unpleasant conditions to Sicily and southern Italy. There are occasions when Saharan winds blow as far north as Britain, where they may cause thunderstorms because of the heat they bring. When it rains, the Saharan dust is washed to the ground as red rain.

site see **position**

situation see **position**

sketch map A rough map drawn to show specific information. Generally not very accurate, it does not need to show a wide range of information, such as might be found in an atlas map.

sketch section A cross-section which is designed to show certain specific information. It does not have to be as accurate as a true cross-section, but will be used to show quite clearly certain features of landscape or perhaps land use.

slack A hollow in between two sand-dunes in a coastal region. The water-table may be near the surface of the slack, and it will be inhabited by very different types of plants and animals compared with those living in the very dry conditions on the sand-dunes.

slag The waste products which result from smelting iron.

slash and burn The method used by shifting cultivators. They slash down the trees and some of the smaller vegetation, and then set fire to it, in order to clear an area for planting crops. Burning the vegetation creates a layer of potash, which is a useful fertilizer. The fire will kill some pests, but it also destroys useful animals and part of the humus layer.

slate Metamorphosed clay or shale. It is normally dark green, black, or blue in colour, and splits easily because it has well developed cleavage. It divides up into smooth pieces, which can be used for roofing material. Many old slate workings have closed, because tiles are now used instead of slate. In north Wales some slate workings have been turned into museums of industrial archaeology, which have become tourist attractions. They make souvenirs out of the slate, such as ash trays, table lamps, small mats, etc. In the Lake District roofing slate was made from the local rock, called Borrowdale volcanics, which is an igneous rock rather than a genuine metamorphic slate.

sleet A combination of rain and snow.

slip-off slope A gentle slope found on the inside of the bend in a meander. It is the slope which leads up to a spur protruding from the side of the valley into the plain. It is on the opposite side of a river from a bluff, and is likely to be an area of deposition rather than erosion. As the river slows on the inside of the bend, deposition of silt or sand may occur, which gradually builds up the slip-off slope.

slump Also **slumping** A type of mass movement in which rock and soil move downhill. Rocks shear away from a hillside or cliff with a rotational movement, leaving a scar on the rock face. Slump is most likely to occur when rocks are tilted, and when impervious clays or shales are overlain by permeable rocks. It has occurred frequently along the Dorset coast, in Kent, near Dover and Folkestone, and in several places along the Cotswolds scarp.

slurry A mass of wet mud, which is likely to slip down a slope. Slurries can be found in shales and some clays along the Dorset coast; they can also occur in man-made conditions, as on dumps and tips of coal waste in mining areas such as south Wales. The Aberfan disaster was the result of slurry rushing downhill into the town.

smallholding A small area of land used, generally intensively, for farming purposes. Smallholdings are often located near towns, and the work is done by the owner and his family, sometimes on a part-time basis, as a second job. Fruit, vegetables and flowers may be grown, and there may be pigs and poultry. Glasshouse cultivation may also take place. Smallholdings are common in many European countries such as Britain, Belgium, the Netherlands and France. In Britain a small-holding is officially defined as a farm of less than 20 hectares.

smelting The production of metal by melting the ore and extracting the impurities. Some mineral ores such as iron may be as much as 65% pure, and so the waste products are less than the weight of mineral extracted. In other cases, such as copper or tin, the ores are less than

10% pure, and may be as low as 1%, and so there are vast quantities of waste products.

smog A mixture of smoke and fog, which used to be very common in London. Since the Clean Air Acts of the 1950s, London has been much cleaner, and smogs are rare. Fogs still occur, but the atmosphere contains fewer pollutants to provide nuclei for the water droplets. Smogs occurred when anticyclonic conditions caused radiation fog to develop. If the air remained calm, the fog could get steadily worse, day after day. More soot and industrial grime would also accumulate, as it was unable to escape into the atmosphere, and pea-soup fogs thickened until visibility was often less than 10 m. Other industrial cities had similar problems, especially if they were in hollows, which accentuated the accumulation of fog and nuclei. Pittsburgh and Sheffield are two notable towns which have been cleaned. One of the largest problems today is in Los Angeles, where the pollution is the result of the large number of motor cars. Los Angeles is in a large basin, noted for inversions of temperature, which create ideal conditions for the formation of fogs and smogs.

snout The downstream end of a glacier. It will be an area of melting for much of the year, and a river is likely to be flowing from underneath the ice, often through a cave. Masses of morainic debris may make the end of the ice look very dirty, and there will be depositional material all around the snout, and also further down the valley, if the glacier is in a period of retreat.

snow A type of precipitation consisting of water vapour which has frozen into ice crystals. Several ice crystals join together to form a snowflake, which will gradually fall down to earth. Some snowflakes may melt as they fall, especially if the temperature is only just down to freezing point. In such conditions sleet occurs, and for the same reason it is possible for clouds to give snowfall on mountains, while producing rainfall on adjacent lowland areas. Snow is melted before being measured in a rain gauge, and on average 10 cm of snow represent 1 cm of rainfall. This varies slightly, depending on the type of snow that has fallen. Some snow can be dry and powdery, if it has come from a dryish area, but other snowfalls can be wetter, if their source is over the sea.

Interior regions, such as the prairies, generally have dry snow, which is blown about in blizzards; it can be moved off roads by a snow blower. Wetter snow occurs in Norway, which requires a snowplough to move it off the roads. In Britain dry and powdery snow sometimes comes from the cold continent in January or February, but wet snow comes from the north or west.

snowfield An accumulation of snow in a mountainous area. If there is enough snow to compress the snowfield it will gradually turn into ice, and eventually glaciers will be able to form.

snowline The line on a mountainside above which there is perpetual snow. There may be snow below the line during the winter months, but it will not persist throughout the year. The altitude of the snowline varies with latitude. At the equator it is about 5000 m; for example, in the Andes or on Mt Kilimanjaro. In the Alps the snowline is about 3000 m, but it varies from north to south and there is further variation depending on the aspect of the slopes. In northern Norway, round the Arctic, and in the Antarctic, the snowline is at sea-level; for this reason glaciers flow into the sea to form icebergs. In Britain, if there were mountains of sufficient height, the snowline would be about 1800 to 2000 m.

softwood Timber which is obtained from coniferous trees. The main use of softwood is for turning into pulp for the manufacture of paper. Some is used as sawn timber for building houses and furniture, and fragments are used in making chipboard. Most of the trees planted by the Forestry Commission and similar organizations in other countries are coniferous. This is partly because of the demand for softwood, but also because softwood trees reach maturity far more quickly than hardwoods. It takes a softwood about 30 years in the British climate, but it will take at least 80 years for a hardwood, such as oak or chestnut.

soil The upper layer of loose material resting on top of the rock which makes up the surface of the earth. It consists of tiny particles derived from the broken-down fragments of rock, together with accumulations of plant remains. The organic remains provide the humus and the inorganic particles provide vital minerals. The depth of soil varies from a few centimetres in some arid regions and on mountainsides to several metres in areas of temperate grassland, temperate deciduous forest and some tropical areas. The soil contains pores, in which air and water can be retained. There are many types of soil, and various methods of classification. A common classification is into zonal, azonal and intrazonal.

soil conservation   The protection and preservation of the soil, so that it will not be ruined or eroded. Soil is slow to form, but if not managed sensibly it can be lost in a very short time. In order to conserve it, it should not be left bare for too long; otherwise, wind and rain may start to erode it. Ploughing, overgrazing or removing trees can expose soil to erosional forces. If soil is used repeatedly for the same crop, certain minerals can be exhausted, and the soil may be ruined. Once it has lost some of its component parts it breaks up more easily and can be washed or blown away. Soil exhaustion is a major contributor to soil erosion. Soil can be conserved by planting grass or trees to protect it, or by planting windbreaks of trees around an exposed area. Run-off should be slowed down, and trees and their roots are vital for this task. In many areas, hillsides are terraced in order to prevent the soil from being washed away down the hillside.

soil creep   The slow downhill movement of soil particles. The movement can take place if the gradient is as slight as 2°. It is generally so slow that it cannot be seen, but over long periods large quantities of soil and rock fragments slide to the bottom of hillsides. The movement is helped by the effects of lubrication, and soil creep is much more active in wet areas. See also **mass movement, solifluxion.**

soil erosion   The removal of soil from an area by means of wind or water. Soil erosion is most likely to occur where the soil has been left bare. This does not happen very often in nature, but it is found in many locations which have been farmed. Erosion may be sheet or gully, and may be caused by wind or water. Wind and water erosion are likely to occur at the same time in the same places. Areas which have suffered from severe erosion include the dust bowl on the High Plains of the United States, many areas in the Sahel, the mountains of southern Italy and the uplands of New Zealand.

soil profile see **profile** (def. 1)

soil texture The size or arrangement of the particles which comprise a soil. They may be fine and small in the case of clays, but are larger and coarse in the case of sands. Often they will be a mixture, called a loam.

solar power Heat, energy, or electricity derived from solar radiation. It is a source of energy which has not yet been exploited to the full, although undoubtedly it will be one day. It is already used in many places on a small scale. There are many houses which have solar panels to create heat for water supplies or central heating. There are solar power irrigation pumps in use in some parts of the Third World. Solar power is most likely to be useful in tropical areas, which have long hours of sunlight. Deserts often average up to 3000 hr sunlight per year, compared with a figure of about 1500 for much of Britain.

solfatara A vent or small crater in the surface of the earth which emits steam and a variety of gases. Many of the gases are sulphurous. Solfataras are thought to be signs of dying volcanic activity, and they occur mostly in volcanic areas. Solfatara (derived from the Italian for sulphur), an area near Naples, gave its name to the phenomenon, and there are other examples in Italy, as well as in Hawaii, New Zealand and elsewhere.

solifluxion Also **solifluction** Soil flow which is slightly faster than soil creep, though still very slow, and not visible to the naked eye. It is the movement of soil and rock particles down slopes as a result of saturation, followed by freezing and thawing. As the water freezes at night and then thaws during the day, particles of soil are moved and then slip downhill because of the effects of gravity. Both water and ice make these movements easier by means of lubrication. Solifluxion is most likely to occur in tundra areas, and was very active in the periglacial conditions experienced in Britain during the ice age. Good examples of solifluxion can be seen on the north side of the South Downs, near Brighton. See also **mass movement, soil creep**.

solstice Either of two dates in the year when the sun appears to stand still. Because of the tilt of the earth's axis and the revolution of the earth around the sun, it appears that the sun is moving. As it moves northwards from the equator in March towards the tropic of Cancer, the Northern Hemisphere experiences its summer. When the sun reaches the tropic of Cancer on June 21st, it appears to stop moving, before returning to the south; this is the summer solstice. Similarly, when it reaches the tropic of Capricorn on December 22nd it apparently stops again, and this is the winter solstice. On the summer solstice the sun is directly overhead at the tropic of Cancer, and it is the longest day of the year for the Northern Hemisphere. At the winter solstice it will be the shortest day of the year.

solution A form of chemical weathering. It is particularly active in limestone areas where the joints can be enlarged to form grykes and pot holes. Solution can also be active in chalk areas and anywhere with rocks containing salts.

South, The The name used by the Brandt Commission to describe the poorer countries of the world, as opposed to the North, which includes Europe, North America and Australasia. Most of the South is in tropical and sub-tropical latitudes, and it includes large areas of South America, Africa and Asia.

sovkhozy (*Russian*) A state farm in the Soviet Union. It differs from a collective in that it would have been set up by the government in an area which had not been fully farmed in the past, or in an area with particular problems. For example many parts of European Russia near to the Baltic have large areas of glacial drift, which are not very fertile. Drainage and reclamation schemes were set up by the government, special equipment was provided, and financial assistance was given to help with the management of the land. There are state farms in the Arctic, where research into growing crops in the very short summers is being conducted. When the dry steppes were ploughed in the 1950s, there were no farms there, and so state farms were created. They had manager and labour supply brought in from other parts of the Soviet Union.

space An area of unspecified size. Spatial distributions are very important in geography. Most studies have to consider the way in which space is utilized. The distances between two locations is also important, as this will influence spatial relationships.

spatial analysis An analysis and explanation of the distribution of people or an activity. In spacial analysis attempts will be made to establish which factors have been instrumental in explaining and accounting for the distribution.

spatial distribution The manner in which settlements or other spatial phenomena are set out and the distances between them. It can be explained in a descriptive way, but it will also need to be quantified if possible, by means of mapping, or by use of the nearest neighbour analysis.

Spearman rank correlation The method used by Spearman to compare a paired set of ranked values in order to look for a correlation. The higher the correlation coefficient, the greater the correlation between the two sets of numbers.

sphere of influence The area around a shop, settlement or other central place, which is affected by the services or goods on offer. In the case of a town, the sphere of influence includes all the surrounding countryside and villages from which customers come to the shops, schools, hospital or the like. A small village has a small sphere of influence, whereas the sphere of influence of a large town such as London may extend for hundreds of miles. See also **central place theory**.

spit A depositional area along a coast, which has been built up as a result of longshore drift. A spit is attached to the land at one end, and may consist of sand, mud, shingle or any combination of these. Most spits protrude across an estuary, and may eventually block the estuary to form a delta. Good examples of spits can be seen at Orford Ness in Suffolk, and Borth in central Wales. See also **bay bar**.

spontaneous settlement A settlement which grows up without planning, and can often appear very quickly, such as a shanty town. At first, the housing is of very poor quality, although this would gradually improve if the settlement became permanent. Most spontaneous settlements are the result of homeless people making use of an empty piece of land, often in a rather undesirable location. Also called **squatter settlement**.

spot height A height marked on an ordnance survey map a point where the surveyor took an accurate measurement. Spot heights are often located on roads or hill tops.

spring A point on the earth's surface at which underground water emerges. It is generally located where there is a change of rock type, and the water flows out at the top of an impervious layer, such as slate, shale, or clay. There are many springs at the foot of limestone and chalk hills. Spring water is often very clean and pure, though it will normally contain large quantities of calcium in solution.

spring equinox see **equinox**

spring line A series of springs along the foot of a chalk or limestone hill, which may give rise to farms or even villages. The spring line generally follows the junction of the limestone or chalk and an impervious rock which makes up the lowland area adjacent to the hills. Very good examples of spring lines can be seen along the north side of the South Downs, or along the south side of the North Downs. There are other good examples at the foot of the Lincoln Wolds.

spring tide see **neap tide**

spur An area of high ground which protrudes into an area of lower ground such as a river valley. Spurs alongside valleys are gradually eroded by the river, but in the upper and middle courses especially, they often form interlocking spurs. In glacial valleys many of the spurs will have been truncated.

squatter settlements see **spontaneous settlement**

stability of air A state of equilibrium of the air in the atmosphere, which occurs when its lapse rate is less than the dry adiabatic lapse rate. This means that if the air is moved upwards, it will be cooler and therefore heavier than the surrounding air, and it will gradually sink down again to its original level. Subsiding or stable air will not give any precipitation. Stable equilibrium can also be achieved when inversion of temperature has taken place. Air in high pressure areas is likely to be stable.

stack* A small rocky islet, generally located at the end of a peninsula. It is usually a result of erosion, which opened up joints in the peninsula to enable caves to form. As the caves grew larger, arches would develop, so that eventually a break could open up in the peninsula. When this happened the end of the peninsula may have been left as an isolated remnant, a stack. There is often more than one stack at the end of a

stack

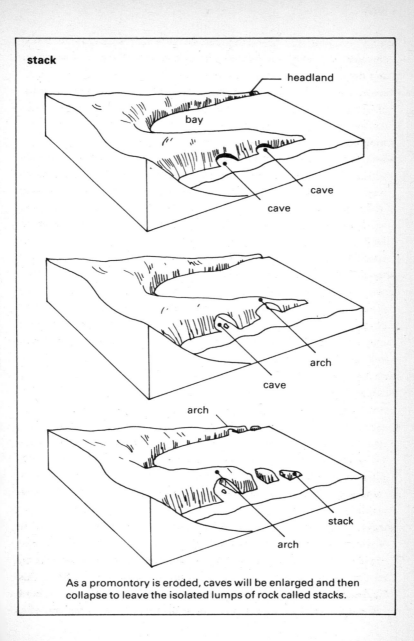

As a promontory is eroded, caves will be enlarged and then collapse to leave the isolated lumps of rock called stacks.

peninsula, evidence of considerable erosion over thousands of years. A famous example is the Needles in the Isle of Wight, and there are examples to be seen on most rocky coasts.

stalactite An icicle-shaped deposit, usually of calcium carbonate, found hanging from the roof in an underground cave. Stalactites are formed by water slowly percolating through the rocks; as the water droplets are about to fall from the cave roof, a tiny layer of calcium precipitates out of the water and solidifies. Gradually, the deposits build up to form a column of calcium carbonate suspended from the roof. Stalactites can be seen in most caves in Carboniferous limestone areas, notably near Cheddar in the Mendips and around Ingleborough in the Pennines. See also **stalagmite**.

stalagmite A column of calcium carbonate which grows upwards from the floor of a cave in a Carboniferous limestone area. It is the result of water dripping from the ceiling of the cave and leaving tiny deposits of calcium on the floor. The deposits gradually build up, and can sometimes reach up to meet the stalacite above. Stalagmites tend to be fatter than stalactites, possibly because the water splashes out when it drops to the floor causing a wider spread of calcium. Stalagmites and stalactites probably grow at a rate of 1 cm per 100 years on average. They can be seen in many caves in the Mendips, at Cheddar and Wookey Hole, as well as in the Peak District in the Pennines, near Buxton, Castleton and Matlock.

staple diet A main and basic foodstuff. In Britain bread is a basic food eaten by everyone, but most people have a fairly varied diet. Many people in southeast Asia will have rice as their staple foodstuff, and in many parts of Africa maize is the main food. Tropical forest inhabitants may eat manioc as their basic food, and some herders may be dependent on milk from their animals and live on a diet of cheese or yogurt.

steppe An area of temperate grassland found in the Soviet Union and in neighbouring parts of southern Europe, in the Danube Valley. The landscape is often very flat and open, and the climate is too dry to support many trees. When ploughed up the grassland can become very rich farm land, especially for growing cereals. Steppes are very similar to prairies, and have a similar continental type of climate. Winters are very cold, ten or more degrees below zero; but the summers are quite warm, 15° to 20°C. Rainfall is quite light, about 500 mm, and falls mainly in summer. The winters are cold and dry, with only light falls of snow.

Stevenson screen A white box on legs, designed to house weather instruments. It stands about 1.3 m above ground level, and has louvred sides to allow free circulation of the air. It is painted white in order to reflect heat. It will contain wet and dry bulb thermometers and probably maximum and minimum thermometers too. It may also contain a thermograph and a barograph. All weather stations have a Stevenson screen, as it is the standard method for obtaining temperatures which are unaffected by sunshine and strong winds.

stock A granite intrusion which is like a small version of a batholith.

Stone Age The period when early humans were using and shaping objects from stone. They often made use of flint, as in East Anglia. The Stone Age can be divided into the Palaeolithic, or Old Stone Age, the Mesolithic, or Middle Stone Age, and the Neolithic, or New Stone Age, as the people gradually improved their implements and techniques. The Stone Age ended in Europe about 2000 B.C; it ended later in North America, and quite recently in parts of New Guinea.

storm A period of strong winds, which is particularly noticeable at sea.

storm surge A rise in the sea to a higher level than expected as a result of strong winds. It is most likely to be associated with the time of high tides, and is likely to cause flooding. It was a storm surge in the North Sea which caused the severe flooding in eastern England and the Netherlands on 31 January 1953.

stratigraphical column see **geological column**

stratosphere The layer of the atmosphere which is above the troposphere and tropopause.

stratocumulus A layer of cloud which also contains some areas with vertical extent. The vertical patches of cumulus may cause heavy showers of rain. If numerous cumulus clouds merge into one large sheet they are known as **cumulostratus**.

stratum A layer of sedimentary rock, usually one of a series which have been deposited one on top of another, as, for example, on the seabed.

Each stratum will indicate the conditions which prevailed at the time of its deposition. The study of the strata is called stratigraphy, and the stratigraphical or geological coloumn contains all the periods of geological time. Rock strata may be very thin, just a few centimetres, or they can be several metres in extent. Initially, they are usually horizontal, but they are likely to be affected by folding and faulting.

stratus A low horizontal sheet of cloud. A thin layer of stratus is light grey in colour and will give no rain. Thicker strati are darker, and may give light or even steady rain. Strati often cover mountains, whilst neighbouring lowlands remain cloud-free. Strati are often found in warm sectors of depressions.

striation The process in which scratches and grooves (striae) are left by ice scraping over rocks. It is the small rock fragments in the ice which probably do most of the damage. The marks show the direction of the ice movement, and can be seen in many places where there has been recent ice activity.

strike The direction of a horizontal line along a rock stratum, which is at right angles to the dip of the rocks. The relationship between the land surface and the contours is similar to that between the dip and the strike.

strip cultivation A method of farming which helps to conserve the soil. On sloping ground, different crops are planted in strips to ensure that the land is not all harvested and bare at the same time. The strips often follow the lines of the contours, and grass and a cereal are frequently alternated down the slope. Strip cultivation is an effective way of preventing the soil being washed away.

strip grazing The sub-dividing of a field into small strips so that only one part is grazed at a time. The animals, generally cattle, eat the grass in one strip, and are then moved on to the next, which will be bounded by an electrified wire fence. As they progress they leave behind rich deposits of manure. Strip grazing is practised in New Zealand, parts of Britain and in the Alps.

subaerial erosion Erosion on the surface of the earth.

subcrustal convection currents Convection currents in the mantle which are responsible for the movement of the plates on the earth's crust. The currents are caused by the heat present within the earth.

subduction The downward movement of rocks and crustal material at a zone of convergence. When two plates are moving together it is inevitable that one goes beneath the other, and this is the zone of subduction. It is normally indicated by a trench, an island arc, or a region of volcanic activity, such as near the Philippines and Japan.

subglacial Beneath a glacier or icesheet.

subsequent stream A stream which is a tributary to the consequent. The consequent stream would have been the main stream to develop on a slope. Subsequent streams often flow at right angles to the consequent, and generally follow the strike of the rocks. They often flow along the softer and weaker rocks, as in the Weald, and may carve larger valleys than the consequent streams. See also **consequent stream**.

subsistence crop A crop grown by a farmer entirely for his own consumption. See also **cash crop**.

subsistence farming Farming in which the produce is consumed by the farmer and his family. The system produces little or no surplus for sale, unlike commercial farming. Subsistence farming is still found in isolated areas in advanced countries, but it is more characteristic of Third World countries.

subsoil The layer which is situated below the topsoil. It may not contain any organic matter, but is likely to include fragments of rock. It is roughly equivalent to the C horizon.

suburb A built-up area on the outskirts of a city, generally consisting of housing estates built in the 1930s or after World War II. Some suburbs may have their own shopping areas, which are mini central business districts.

summer solstice June 21st. See also **solstice**.

sunshine recorder An instrument for measuring the amount of the sunshine. The standard sunshine recorder is the Campbell-Stokes recorder. It consists of a glass sphere, which concentrates the rays of the sun and focuses them onto a piece of special graduated paper. The sun makes scorch marks on the paper, and this can be measured to give the length of time that the sun was shining.

supercooled A term used to denote drops of water which have been cooled below freezing point but have remained in liquid form. Super-cooling can occur if the air is very still. The supercooled droplets freeze onto objects moving through them, such as the leading edge of aircraft wings, and if they drift with a very gentle air movement they may freeze onto the sides of buildings, or trees, or telegraph poles.

superimposed A term used to denote a river system which does not follow the trend of the rocks over which it is flowing; this is because it is continuing to follow the trend of other rocks which have since been eroded.

surface run-off The portion of rainfall which runs off either as channel flow or as overland flow, but does not percolate into the soil and rocks, and is not evaporated into the atmosphere. Also called **immediate run-off**. See also **delayed run-off**.

surge A large or massive movement of water, generally in the sea. See **storm surge**.

suspension The state in which small particles of sediment are carried by rivers. The particles are held up in the water by turbulent upward eddies, which prevent the tiny particles from falling to the bed of the river. It is the suspended particles which make rivers look muddy. It has been estimated that in a year 80 tonnes of solid matter are removed from each square kilometre of the earth's surface; so there is a lot of material for the rivers to carry. The Mississippi carries 300 to 400 million tonnes in suspension each year.

swallow-hole A large pothole in a limestone area down which a river is flowing. The hole will be an enlarged joint, which has been weathered by solution and also eroded by running water. Some swallow-holes descend for up to 100 m; for example, Gaping Gill on Ingleborough in the Pennines. See also **sinkhole**.

swash see under **backwash**

syncline The downfold in folded strata. The upfold is called the anticline. The rocks in the syncline are often compressed, which makes them harder and more resistant to erosion, and they often erode more slowly than the rocks in the anticline. See also **inverted relief**.

synoptic chart A synopsis or summary of the weather which is drawn on a map each day. The chart shows the exact meteorological conditions at a certain time, with isobars, weather systems, temperatures, wind speeds and directions, amounts of cloud, etc.

synthetic fibre A fibre which does not occur naturally and is manufactured by chemical synthesis from wood, coal or petrol. The earliest synthetic was nylon, made from coal and developed commercially in the 1940s. In the 1950s rayon was made from wood, and since then there have been many discoveries and developments, mostly using petroleum as the raw material. Terylene, Acrilan, and many others are available. They have special qualities; they may be drip-dry, crease-resistant and hard-wearing, but none of these synthetics has quite replaced all the qualities of wool and cotton. Also called **artificial fibre, man-made fibre**.

systematic geography That part of geography which deals with the overall study of a particular subject, such as agriculture, transport or industry, rather than the study of these subjects within a particular region. It is an approach which gives a fuller understanding of the topic or system under consideration. For example a glacial system is a glacier, together with all its inputs and outputs. If one specific glacier has been studied, it is likely that other glaciers will be understood because of the similarities. Any geographical subject can be considered as a system. For example, an urban system; the study of one town in detail, will help to explain how all towns have developed and function.

T

Taafe A model named after E.J. Taafe, who made a study of the growth of communications in an African country. There are various stages of transport development. Each tiny port has a small hinterland served by paths and trackways. A road or railway may be built to link an inland mine or estate with the port. Inland centres which grow quickly will be linked with other inland towns, and the major port or ports will gradually build better roads or railways to the inland towns.

taiga The coniferous forest which extends over thousands of square kilometres in Siberia, Russia, Finland, Sweden, Norway and Canada. The climatic conditions are unsuitable for deciduous woodland, because the growing season is seven months or less, but coniferous trees can survive. Summer temperatures are around 15°C, and January averages −10° to −20°C. There are several months with temperatures below zero, and the subsoil is frozen for much of the year. The roots of the coniferous trees spread out horizontally as they cannot penetrate vertically into the ground. The main types of tree are spruce, larch, fir and pine. In the southern parts of the taiga, the trees grow to 15 m in height, but further north the trees become smaller until they are no more than bushes, and tundra vegetation can be seen. The trees are a source of softwood for pulp, which is used to make paper. Thousands of trees are cut down every day, but in most countries reafforestation ensures that the trees are replaced. Many highlands such as the Alps have coniferous forests which are similar to the taiga. Wherever coniferous forests occur, the needles fall to the ground to produce podsols, which are acid and ashy grey in colour. See also **coniferous forest, deciduous woodland, evergreen trees**.

take-off point Stage 3 of **Rostow's model**. All advanced countries have gone through the various stages of the Rostow model, and take-off began at different times in different countries. In Britain take-off occurred about 1800, in France and the United States it was about 1860, and in Japan about 1900. In Brazil it occurred in the 1950s.

talus An accumulation of small rock fragments which formed on the side of a hill and then slid down to its base, because of the effects of gravity and lubrication by water. The rock fragments are the result of weathering, especially freeze-thaw activity, and are generally referred to as scree. Talus commonly consists of angular fragments of rock, which if they were cemented and consolidated would be called breccia. When they slide downhill they may accumulate on the valley floor in a triangular or cone shape, rather like a rock delta. All types of rock can form talus and good examples can be seen on granite areas of Dartmoor and the limestone areas of the Pennines.

tank A lake formed by building a dam across a stream or an embankment around a field which has been covered by flood water. Water collects in the tank during the rainy season, and can be used for growing crops when the rainy season has finished. Such supplies of water will not last throughout the year, but they are retained for a few extra weeks. As the water level subsides, crops are planted in the saturated soil. Tanks are found in many parts of India.

tanker A ship, aircraft, road or rail vehicle for carrying fluids, especially oil, in bulk. Road tankers carry oil from refineries to consumers, and take milk from farms to dairies. In the 1950s seagoing tankers were normally 20–30,000-tonne vessels, but throughout the 1960s the size of tankers increased until 300,000 tonners were in use. The large tankers were able to transport oil more economically than the smaller vessels. Because they need very deep water in the harbour, special oil ports grew up to handle the tankers, for example, Milford Haven and Europort.

tarn A small lake, especially in the Lake District. Tarns are often the lakes which have formed in corries; for example, Red Tarn on Helvellyn. The name is derived from the old Norse word for a teardrop.

tear fault A fault in which the movement is along the horizontal. Tear faults occur frequently in areas of tectonic activity. The amount of movement may only be a few metres, but can be several kilometres in some cases. For example, on the Caledonian Canal fault in Scotland the granite found at Strontian, southwest of Ben Nevis, used to be alongside the Foyers granite. They were both part of a granite mass formed in the Caledonian orogenesis. San Andreas in California is a well-known example of a tear fault. Also called **lateral fault, wrench fault**.

tectonics The movements which affect the features of the earth's crust. The processes which bend or crack the crustal rocks to cause folding or faulting are the result of tectonics, resulting from movements of the earth's plates.

temperate Having a moderate climate, such as mid-latitude areas, including Britain. Strictly speaking the temperate zone is the area between the tropical or torrid zone, and the frigid zone; it extends from the tropic of Cancer to the Arctic Circle and from the tropic of Capricorn to the Antarctic Circle. The summers in temperate latitides tend to be warm, but the winters are cool. Temperate climates are those which have neither very high nor very low temperatures.

temperature Temperature is measured by a thermometer and generally recorded in degrees centigrade; 0° is the freezing point of water and 100°C is the boiling point of water. The sun is the major source of heat. Temperatures generally decrease from the equator towards the poles because of the corresponding decrease in the angle of the sun away from the equator. Temperatures also decrease with height above sea-level because the air becomes thinner. Latitude and altitude have both been mentioned as major factors influencing temperature, and the third factor is the difference between land and sea. Land heats up and cools down more quickly than sea, and so it is generally hotter in summer and colder in winter.

temperature inversion An anomalous increase in temperature with height. Normally the temperature of the air decreases from ground level upwards. The average rate of decrease is 1°C for every 160 m. In certain meteorological conditions the normal situation is reversed. On a clear, calm anticyclonic night, the cool air may roll downhill and accumulate in valleys, and the air temperature will be lower near the valley bottom than it is 100 to 200 m higher. Above the cold layer there will be warmer air, which is likely to form cloud or haze. On summer evenings it is often possible to see evidence of a warmer layer, if there is smoke rising from a bonfire. The smoke will rise vertically and then bend horizontally when it reaches the **inversion layer**. If this situation develops on a larger scale, the dust and dirt rising into the atmosphere are trapped and unable to escape, giving rise to serious pollution, as for example, in Los Angeles. As the weather conditions are often anti-cyclonic and Los Angeles is situated in a hollow, the fumes from cars cannot escape very easily, until the air becomes more turbulent causing the layers to mix and the inversion layer to disperse.

tephra   The ash, dust and cinders thrown out by a volcanic eruption. The term does not include lava. When tephra is thrown up into the air by an explosion, it will soon fall again, quite close to the crater. The smaller and lighter particles are blown further away and accumulate on the downwind side of the eruption. Large deposits of tephra were left by the eruption of Heimaey in Iceland in 1973. So much tephra fell on to some houses that the roofs collapsed because of the weight.

terminal moraine   A moraine formed at the end of a glacier when the ice has reached its maximum extent and begins to melt. Terminal moraines are generally slightly curved or arcuate in shape. Because of the changing location of the snout several **recessional moraines** may form at the end of a glacier, and if the ice is thick, they can form small ranges of hills. For example, the Baltic Heights in north Germany reach over 300 m, and the Cromer Ridge in north Norfolk is over 30 m. Part of the city of York stands on a terminal moraine.

terracette   see **sheep track**

terracing   A series of terraces cut into a slope in order to create strips of flat land. Land is dug out of the slopes and used to build up new flat areas. Walls or mud banks are used to preserve the level patches and to help retain water if irrigation water is being supplied to the terraces. Terracing is used for farming in a variety of regions; for example, in rice-growing areas in southeast Asia and in wine-growing areas of France. It was also used by the Incas for producing maize on the Peruvian Andes.

terra rossa   (*Italian*) Red earth which has gained its colour because of the iron content. Terra rossa is a residual soil found in limestone areas of Yugoslavia.

terra roxa (*Portuguese*) A reddish iron-rich soil which formed on the basaltic rocks of the São Paulo state on the Brazilian plateau. It is particularly suitable for growing coffee.

Tertiary The geological era which followed the Primary and Secondary, and preceded the Quaternary. The Tertiary is the era of mammals, when modern animals and also modern plants became dominant throughout the world. It began about 70 million years ago, and includes the following geological periods: Palaeocene, Eocene, Miocene, Oligocene and Pliocene. The Alpine earth movements occurred during the Tertiary, which formed all the young fold mountains, such as the Alps, Himalayas, Andes and Rockies. Most of what is now Britain had become land by the beginning of the Tertiary, but there was still sea in the southeast. Some new sedimentary rocks were formed in the London basin and Hampshire, including the Isle of Wight. There was volcanic activity in southwest Scotland and northeast Ireland, caused by fissure eruptions where North America was drifting away from Europe.

tertiary sector The part of the economy which includes all the service industries; that is, the jobs connected with administration, transport, education, medicine, banking, etc. In highly developed countries, such as those of western Europe or North America, tertiary industries often employ over 50% of the work force, especially in the large towns.

textile Wool, cotton, flax and silk are sources of natural textiles, and there are many synthetic fibres such as nylon, rayon, acrilan and terylene. The manufacture of cloth has seen a recent decline of natural fibres as synthetics have become cheaper and more versatile. The world's major industrial nations, such as Japan, the United States, Korea and Taiwan, are the leading suppliers of textiles.

thalweg Also (*German*) **talweg** see **long profile**

thermal electricity Electricity produced from coal, oil or natural gas. The raw materials are burnt to produce steam, which drives the turbines. Coal-mining areas, such as the Trent Valley in England, or West Virginia in the United States, have several thermal power stations; the same is true of oil refining locations, such as Middlesbrough on Teesside or Houston in Texas.

thermograph An instrument which keeps a written record of temperature. It consists of a thermometer made up of two strips of metal, which expand and contract with changes of temperature, and this is attached to an arm with a pen, which records on to a special piece of paper attached to a revolving drum. A clockwork mechanism inside this drum turns it round once in a week. At the end of the week the paper can be taken off, and it will show the temperature changes of the previous week. A thermograph is not as precise as a thermometer, but it shows all the changes and trends quite clearly.

thermometer An instrument for measuring temperature. There are various types such as maximum, minimum, grass minimum, and wet and dry bulb. Some contain mercury, and others contain alcohol; they both work efficiently. Temperatures can be read off a scale on the glass tube.

Third World Those countries, especially in Africa, Asia and Latin America, which are aligned with neither the developed nations of the west nor with the members of the Communist bloc. The expression is rather unsatisfactory as it includes countries which are quite different from one another in many respects.

thorn forest A type of thorny scrub found in areas with prolonged periods of dry weather. In order to survive the periods of drought, the plants often have thick bark, or long roots, and are covered with thorns.

threshold 1 See **sill** (def. 1). 2 The end of a fjord at the point where it enters the open ocean.

threshold population The minimum number of people in a region required to support a particular shop or service.

throw see **fault**

thunder The loud noise which accompanies a flash of lightning and is caused by violent disturbance of the air by an electrical discharge. Because the speed of light is faster than the speed of sound, the lightning appears to precede the thunder, unless it is directly overhead. The delay between lightning and thunder is a rough indication of how far away the storm is.

thunderstorm A storm caused by convectional activity and accompanied by thunder and lightning. Because of the size of the convection currents, some thunderstorms give very heavy rain, sometimes several centimetres in an hour, and can cause serious flooding. Equatorial areas receive convection rain and may have a thunderstorm nearly every day. In Britain and other parts of Europe there are thunderstorms in the summer months or early autumn, when the land is hottest and strong convection currents are likely to form. Prolonged thunderstorms develop when there is a large supply of moisture, so that giant cumulonimbus clouds can form, and a high lapse rate, which will ensure a rapid and persistent upward movement of air for 3–4000 m above the base of the cloud. If the drops of water are forced so high that they freeze, there will be hail as well as heavy rain. See also **convection rainfall, lightning, thunder.**

tidal limit The highest point in a river inlet which is ever reached by the sea.

tidal range The difference between the water-level at high tide and the water-level at low tide. In some places in the Mediterranean, the tidal range is only a few centimetres, whereas at Avonmouth on the Severn estuary the range is as much as 10 m when there are spring tides. When there is a very high spring tide, the tidal bore rushes up the River Severn as far as Gloucester. The Bay of Fundy in eastern Canada has a tidal range of 15 m.

tidal wave see **earthquake**

tides The periodic rise and fall of the sea caused by the pull exerted on earth by the moon, and to a lesser extent, by the sun. In many parts of the world there are two high tides and two low tides every day. The time of each high tide is 12 hr 20–25 min later than the preceding tide, because the position of the moon relative to the earth will have changed by a small amount after 12 hours have elapsed. See also **neap tide, spring tide.**

tidewater (Of a location) situated close to tidal water, either on the seashore or in a tidal estuary. Tidal water enables bulk-carrying vessels, the cheapest form of transport, to penetrate further inland; as for example, in London.

tierra caliente (*Spanish*) The hottest parts of tropical Central and South America, which are the low slopes of the Andes found near the equator – on the coastal plains and in the lower Andes at altitudes of up to about

1000 m. In the tierra caliente tropical forests can grow because there is a constant temperature of about 25°C and up to 2000 mm of rainfall spread right through the year. Manioc is the major food crop, and bananas, sugar cane and cocoa are grown.

tierra fria (*Spanish*) The cold areas of tropical Central and South America, which are found higher up in the Andes, at about 2000 to 3000 m in equatorial latitudes. The average monthly temperature is about 15°C throughout the year. The total rainfall is spread throughout the year but is no more than 1000 mm. The rainfall in equatorial latitudes is convectional, which is caused by heat. As the temperature decreases with height, there is less heat and therefore less convection. For this reason rainfall decreases with height in equatorial areas, in contrast to temperate latitudes, where rainfall increases with height. In the tierra fria wheat is grown and barley, too. Potatoes are a staple food and also a source of alcohol. Sheep and llamas are reared. The natural vegetation is coniferous forest in the lower parts, with scrub and grassland higher up. In the tierra fria climatic conditions are more suitable for Europeans than in the tierra caliente, and therefore several large Andean cities have grown up at high altitudes; for example, Quito, the capital of Ecuador, La Paz, the capital of Bolivia, Caracas, the capital of Venezuela and Mexico City.

tierra templada (*Spanish*) The temperate areas of tropical Central and South America, which occur between 1000 m and 2000 m in the Andes near the equator. Temperatures average 20°C each month, and the total rainfall is about 1500 m. Wheat and maize are grown as food crops; and coffee, cotton and a variety of fruits are grown as commercial crops. Much of the land is naturally forested, but when cleared it is suitable for agriculture. The original Indian inhabitants, the Incas, made terraces on many mountain slopes in the templada as well as in the fria zone.

till Glacial deposition; rocks, sand and fine-grained clay can all be classed as till. Some deposits are very sandy, whilst others consist mainly of clay, and some, as in Norfolk, are very calcareous because the icesheet passed over a large area of chalk. The northern part of Long Island New York is till, and the southern part of the island is glacial outwash, separated from the till by a terminal moraine. Till is generally unstratified. Also called **boulder clay**.

tombolo A bar or spit which is joined to an island. Tombolos are not common, but Chesil Beach in Dorset is a very good example. The shingle ridge of Chesil is joined to the Isle of Portland.

topography The relief and configuration of a landscape, including its natural and man-made features. The hills, valleys, forests, fields, roads and settlements are part of the topography. A topographic map endeavours to show all such features, and needs to be quite large scale to show detailed information. The ordnance survey 1:50,000 series is an outstanding example of this type of map.

topological map A map in which lines are stretched or straightened for the sake of clarity, but without losing their essential geometric relationships. A famous topological map is the plan of the London Underground. Sometimes called rubber sheet geometry, the technique can be used for any route map and the end product will be simplified and clear.

topsoil The top layer or A horizon of the soil profile. It contains fine particles, humus, nutrient and the roots of plants. In areas of arable farming it will be cultivated, and turned over by ploughing.

tor An outcrop of granite which stands above the general level of the landscape, as on Dartmoor, Bodmin Moor, and many places in southern Africa. It is an isolated mass of rock, showing vertical as well as horizontal lines of weakness where weathering has attacked the rock. It is a residual rock, the remnant of a layer of granite that has been largely worn away. It has survived either because it was harder than the surrounding rocks, or, more likely, because it was an area in which the joints were more widely spaced and therefore there was a slower rate of weathering. Around tors there are generally masses of weathered fragments, which are referred to as clitter.

tornado 1 A small and intense atmospheric disturbance which travels across the countryside at 10 to 100 km per hr and contains winds of up to 300 km per hr. Tornadoes are associated with intense heating in continental areas in late summer, when land masses are at their hottest. The centre of a tornado is an area of low pressure which sucks up dust to give a blackish funnel rising to the sky. It is claimed that on many occasions cattle and humans have been sucked up by the funnel. Around this funnel of rising air are very strong winds which destroy crops and sometimes buildings. The low air pressure in the centre of a tornado sometimes causes the walls of buildings to fall outwards, because pressure is higher inside than out. Tornadoes are commonest in the interior of the United States, but also occur in India and many other countries. A few small examples occur in England during most summers. When tornadoes cross over water they become waterspouts. The average life of a tornado or a waterspout is 4 min. Also called **twister** (in North America). 2 A violent storm at the beginning of the rainy season in West Africa. They bring strong winds and torrential rain, which is caused by mild air coming in from the sea and meeting the dry north-easterly air from the Sahara.

tourism One of the major growth industries, tourism is important in many developed countries and is possibly vital to the economic development of poorer countries. In Britain London, coastal locations and scenic areas such as the Lake District are important tourist areas. Southern Spain has developed rapidly since 1960 because of tourism, and Tunisia, Gambia, and many less developed countries are now earning money from tourism. The attraction of sunshine is helping economic development in tropical and sub-tropical areas. Tourism has also grown into a major industry in snow-covered regions such as the Alps, though even Scotland has managed to develop a skiing industry.

townscape The landscape of an urban area. It can be dominated by concrete and brick, but may also contain parkland. American cities have skyscrapers near their centres, but their suburbs may spread out over great areas, especially in the case of Los Angeles. European cities are less spread out, and contain large areas of suburban development. In New Zealand, where most residences are single storey, urban areas spread out over fairly large distances.

trade Since spices were first brought to Europe from the Indies, international trade has grown steadily. Large quantities of wheat and cotton crossed the Atlantic from the United States to Europe, and manufactured goods were sent from Europe to all parts of the world. In the late 20th century coffee goes from Brazil to the United States and

Europe, and oil from the Middle East is exported to Japan and Europe. Iron ore from Australia goes to Japan. The wealthy industrial countries often buy raw materials from the less wealthy ones, but all countries are now dependent on trading links with other nations.

trade wind A wind which blows from the tropical high pressure zones (horse latitudes) to the equatorial low pressure zones (doldrums). In the northern hemisphere the air moving from the tropic of Cancer towards the equator is deflected to its right to make it a north-easterly wind. In the southern hemisphere the air moving from the tropic of Capricorn to the equator is turned to its left to become a south-easterly wind. Trade winds are fairly persistent winds; they are all easterlies and tropical. They blow from deserts towards the tropical forest zone.

trading estate A purpose-built industrial site generally on the edge of a town, where good road links, such as a bypass or ring road are available, together with water, gas and electricity. Small factories or workshops are constructed for firms to rent, often at reduced rates because of government grants. The buildings are generally single storey and the industries light. Most trading estates contain a variety of industries which employ both male and female labour. Some of the earliest examples were set up in south Wales in the 1930s, in an attempt to provide employment to offset the loss of jobs in coal-mining and steelworking. Slough developed many industries on its new trading estates in the 1950s, and new towns such as Bracknell, Milton Keynes and Crawley have several trading estates.

transect A line which cuts across. Transect lines are used in field studies; for example, to make a study of sand-dunes, a transect line would be followed inland, across the dunes, and at right angles to the sea. Along the slope of the sand-dunes the changes of vegetation would be observed. The information gathered could then be mapped and a cross-section drawn along the transect line. A transect does not give as much detail as a complete study of an area, but it provides a sample without the necessity of too much time being spent on the task.

transhumance The seasonal transfer of livestock between mountain and lowland pasture. In the Alps, for example, cattle, and sometimes sheep, are taken up the hills to high level pastures above the tree line during the summer months. Some members of the village community stay on the mountains in summer huts to look after the cattle, milk them and make butter or cheese. Meanwhile, down in the valley, other people cut the hay and look after any other crops. When the high-level pastures receive their first snowfalls in autumn, the cattle are moved back to the valley. The same movement of cattle occurs in the fjordland region of Norway, and to a lesser extent in Wales, Scotland and the Lake District.

transition zone   An area in an inner city where the older residential and industrial zones have decayed and are gradually being replaced by new developments.

transpiration   The process by which plants give off and lose water vapour. It passes through stomata in the leaves, and goes out into the atmosphere. From the atmosphere the moisture may be recycled as rain, and then taken up into the plants again, via the roots. Transpiration contributes to the formation of rain. It has been discovered on the edges of the Sahara in Algeria and also in the dry steppes in the Soviet Union, where millions of trees have been planted, that rainfall is heavier than it used to be. Conversely in parts of the Amazon basin where forests have been totally removed the annual rainfall has fallen from 2000 mm per annum to nearer 1000 mm.

transport network   see **network**

transverse coast   see **discordant coastline**

tree line   1 The line on a mountain above which trees do not grow. This is mainly because the temperatures are too low, but the quantity of rainfall or the lack of soil may also be influential. In the Andes, near the equator, the tree line is just below 4000 m, and in the Alps it is about 3000 m. The tree line is generally 1000 to 2000 m below the snowline in temperate and tropical latitudes, though it will be much less in the higher latitudes. 2 A line of latitude beyond which trees do not grow, such as the northern edge of the taiga, where the forest gradually deteriorates and becomes tundra. 3 The lower edge of a forest below which the hillside is cultivated or covered with grass, as, for example, in many Mediterranean countries.

trellis work drainage   see **grid iron drainage**

tributary A subsidiary stream or river which flows into a larger river. Some rivers may have hundreds of tributaries, all of which collect the rain and ground water from a part of the river basin, but rivers which flow through deserts may have few tributaries; for example, the Nile. See also **distributary**.

tropical cyclone The type of violent cyclone characteristic of tropical areas in which wind speeds are likely to exceed 150 km per hr and several centimetres of rain may fall. Tropical cyclones do not occur on the equator because there is too little air circulation; they occur mostly 10° to 20° to the north or south. They form regularly in the Bay of Bengal, causing flooding and devastation in Bangladesh. See also **hurricane, typhoon**.

tropical grassland Grassland found in tropical areas which do not have rainfall throughout the year, but only in the summer months. The total amount may be as much as 1000 mm, but even so, forests will not grow because of the prolonged droughts of the winter months. A few xerophytic trees can survive, including several varieties of acacia, but the main natural vegetation is grass, which may grow to 2 m in height during the hot wet summer, but in winter it will go brown and shrivel up in temperatures of 15° to 20°C. Pastoralists have to walk many kilometres in the dry winter to find pasture and water for their animals. Tropical grassland is found in Kenya, northern Nigeria, northern Australia, the Brazilian plateau and the Orinoco basin. Most tropical grasslands are found between 5° and 20° north and south of the equator, but in East Africa they are also found on the equator, because the land is too high (and therefore cool) to support forest.

tropical rain forest Luxuriant evergreen forest which grows in equatorial lands, mostly in the areas between 5° north and 5° south of the equator. Rain falls most days of the year, and there is constant heat, so growth is rapid. Temperatures are from 25° to 27°C every month, and there is up to 2000 mm rainfall each year. The Amazon basin and Zaïre contain the largest expanses of tropical rain forest, but there are some smaller patches on the coasts of Colombia and northern Ecuador, northeast Brazil, the lowlands of Panama and Costa Rica, and the northern coastal region in Mozambique. Soils are generally poor and shallow in tropical forests. The nutrient for the trees comes from the dead leaves which fall throughout the year. The roots of the trees are quite shallow, and the larger trees have buttresses, which enable them to stand up straight and tall, often reaching a height of more than 30 m. Beneath the taller trees are a layer of smaller trees, which grow to 15 m and beneath them are the small trees of 5 to 10 m. The trees form a dense canopy and very little sunlight can penetrate through to the ground, so that there is little or no undergrowth in tropical forests. The ground has millions of ants and other insects, and there is intense bacterial activity breaking down the fallen leaves, but there are few animals apart from the birds and monkeys found in the trees. There are many lianes, which are vine-like plants, and there are also epiphytes, which are parasitic plants found on tree trunks and branches. Also called **equatorial rain forest**.

tropic of Cancer An imaginary line of latitude at 23½° north of the equator. It is the most northerly latitude reached by the overhead sun, on June 21st, which is the summer solstice. It is because the tilt of the earth's axis is 23½°, that the sun reaches this latitude on its northward movement. The tropic of Cancer passes through Mexico, Saudi Arabia, India and southern China.

tropic of Capricorn An imaginary line of latitude at 23½° south of the equator. It is the most southerly latitude reached by the overhead sun, on December 22nd, the winter solstice. The line of latitude passes through Paraguay, southern Brazil, Botswana and Australia.

tropopause The boundary between the troposphere and stratosphere. The height varies but it is about 16 km at the equator and about 8 km at the poles.

troposphere The lowest part of the atmosphere, below the tropopause. The troposphere contains the water vapour and hence all the weather phenomena of the atmosphere. In the troposphere there is generally a steady fall of temperature with an increase in height; this is called the **lapse rate**.

trough 1 A low pressure area which is an elongation or extension of a depression. A trough of low pressure extends to the south side of depressions in Western Europe, and often contains a cold or occluded front. Troughs may bring rain because of the existence of the front, but they generally pass overhead within 12 hours. 2 A deep trench on the ocean floor; for example, the Puerto Rico Trough or Mindanao Trench. 3 A deep valley, such as a glacial U-shaped valley.

truck farming see **horticulture**.

truncated spur A spur which has been cut off by glacial erosion. Truncated spurs are found in glacial U-shaped valleys and the truncation leaves very steep valley walls. In some locations there are waterfalls, where streams flow over the edge of the valley. At the foot of truncated spurs there are likely to be accumulations of scree. Nant Ffrancon in north Wales and many other U-shaped valleys have truncated spurs, but much larger examples can be seen in Lauterbrunnen in Switzerland and Yosemite in California.

tsetse Any of several blood-sucking African flies which transmit diseases to cattle and humans. To humans the tsetse passes on sleeping sickness, which produces lethargy and an inability to do any work; to animals it transmits nagana. The tsetse occurs widely in tropical Africa, but needs woodlands for breeding purposes. The open grassy plains are better areas for cattle than clearings in the forests.

tsunami see **earthquake**

tuff A rock which consists of consolidated ash and dust thrown out by a volcanic eruption. It is a soft and quite porous rock.

tumulus A mound built at the side of an ancient burial ground. The mound may be 10 to 20 m in height and 20 to 50 m in length. Tumuli date from pre-Roman times. Also called **barrow**.

tundra The sub-arctic area found to the north of taiga regions in Alaska, Canada, Norway, Sweden, Finland and the Soviet Union. Summer days are quite long and often sunny, but average temperatures for July do not rise above 10°C. For 6 to 9 months the average temperature is below zero, and there is a thick layer of permafrost, of which only the top few centimetres thaw out in the summer. Where melting occurs the land is wet and marshy, an ideal breeding ground for mosquitoes. The wet tundra soils are gleys (glei), and waterlogging is characteristic. Generally, trees cannot grow in the tundra because of the unfavourable conditions, but there are some dwarf willow and birch in sheltered hollows. Mosses and lichens are common, and there are many small flowering plants in the summer. Reindeer (called caribou in North America), arctic fox, arctic hare, and many geese, ducks and waders are present in the summer, but most move south in winter. There is no real tundra in the southern hemisphere, though something similar is found in South Georgia and other Antarctic islands. High mountainous areas in the Himalayas, Andes, Norway, and elsewhere also have areas which are similar to the Arctic tundra.

turbulence Irregular movements of the air in the lowest layers of the atmosphere. Turbulent activity causes mixing of the air, and often causes condensation as the rising air cools. Turbulence can cause showers of rain, as well as bumpy conditions for aircraft.

twilight The period of faint light before the sun has risen above the horizon in the morning and after the sun has descended below the horizon in the evening. It lasts only a few minutes near the equator, but can last for an hour or more in the Arctic and Antarctic.

twilight zone An urban area where decline and decay have taken place, and there has been no redevelopment. It is often near the central business district, where old housing is in need of replacement.

twister A North American name for a **tornado.**

typhoon A tropical cyclone which affects the coasts of China, the Philippine, and Hong Kong, generally in late summer or early autumn. Typhoons form over the China Sea as a result of excessive heating, and travel westwards over the land, causing widespread destruction; they weaken once they move inland.

U

ubac The shady side of a valley, normally the north-facing side. It receives sunshine during the summer months only, and is likely to be forested or covered wih grassland. See also **adret**.

Ullman, E.R., and Harris, C.D. Two American geographers who first described the multiple nuclei model of urban areas. They assumed that large cities have more than one central point or nucleus, and that growth spreads outwards from separate nuclei to merge in one large urban area. They believed that urban expansion would swallow up nearby villages, which would then create mini central business districts. Ullman and Harris improved on the Burgess model and Hoyt's sector model, but added a certain complexity.

underdeveloped (Of a country) having little or no industrial development. The term is most frequently applied to Third World countries, where most people are still involved in agriculture of a subsistence type. A low average gross national product is possibly an indication of underdevelopment.

underground stream A stream which runs below the surface of the ground over all or part of the course. In some limestone areas streams travel many miles underground, and such areas often contain potholes, underground channels and caves, in which stalactites and stalagmites form. Because the rivers change course from one bedding plane to another, there are many caves which no longer have rivers flowing through them; for example, Fell Beck flows off Ingleborough in Yorkshire and falls 100 m down Gaping Gill, a large sink hole. It travels underground for 2 km before emerging at Ingleton caves near Clapham. Another well-known location of the emergence of an underground stream is at Wookey Hole, where the River Axe comes from beneath the Mendips. This type of spring is called **vauclusian**.

undernourishment The condition arising from an insufficient intake of food to maintain normal health. Millions of people in Latin America, Africa and Asia suffer from both undernourishment and malnutrition.

urban Relating to towns or cities. An urban area can be defined as a built-up area, according to the United Kingdom census, it means that there is a population of at least 25 people per hectare (10 per acre). In Denmark any settlement with more than 2000 inhabitants is classified as an urban area, and in India it is any settlement with more than 5000 inhabitants. Any urban area should be multifunctional, and not just involved with agriculture, otherwise, it would be called a rural settlement. There is no real distinction between a town and a city, except that a city is likely to be larger. In Britain a city is sometimes defined as a town with a cathedral, but this would include a number of small settlements such as St Davids in Wales, and so it is an unacceptable definition.

urban decay Deterioration and decay, especially of the older parts of a town. It is possible to restore or renovate these areas, and the older industrial towns of England, Wales, Belgium, Germany, and elsewhere have industrial buildings, canals and derelict housing, which are gradually being replaced. The areas suffering from greatest decay are in regions where factories have been closed and become derelict, and housing is sub-standard for the 20th century.

urban field The area around the city from which customers are drawn into the shops. A small town has a small urban field, extending for a few kilometres, but bigger towns such as Cambridge or Guildford have a field extending as far as 30 kilometres. The distance will not be the same in all directions, as it will depend on the availability of roads or railways, and also the proximity of other towns, which may draw customers in a different direction. The urban field of big cities such as London or Paris can extend for hundreds of kilometres.

urban heat island A large town or city which is warmer than the surrounding rural areas as a result of the heat stored by the buildings and derived from the sun, and also as a result of central heating. An isotherm map will illustrate the phenomenon. In London, for example, the inner parts of the city may be 4°C warmer than the surrounding rural areas, and outer parts will be 1° to 3°C warmer. The higher temperatures mean that London has fewer frosts than most of southern and eastern England, and there is less snow too.

urbanization 1 The growth and expansion of urban development and urban areas. 2 The movement of people into urban areas. For example 92% of the population of the United Kingdom is classified as urban. Figures for some other countries are Australia 90%, United States 79%, India 24%, Bangladesh 14%, Brazil 71%, Mali 23% and Kenya 17%.

urban morphology The patterns of land use and the shapes of different zones which have evolved in many cities. It is possible to draw land-use zones for any urban area. There are three standard models of urban morphology, named after the researchers who described them: Burgess, Hoyt, and Harris and Ullman. However, every city is unique and will have its own distinctive features, even though there will be similarities with other cities.

urban renewal The replacement or renovation of declining older urban areas. Urban renewal is a slow and expensive process, but some towns, for example, Rotterdam and Düsseldorf, which were badly damaged in World War II, had opportunities for large scale renewal schemes. A smaller area of urban renewal after World War II was in Coventry, where shopping precincts and a new cathedral were built.

urban sprawl The spreading out of urban areas into surrounding rural areas as new houses are required for the growing population. The growth of the suburbs has been helped by the increasing number of cars, which has enabled people to live further away from their work in the city centre. Before World War II it was the buses and trams which enabled the suburbs to spread out further from the city centre. The inter-war suburbs built in the 1920s and 1930s were generally along the main roads, sometimes as a form of ribbon development.

U-shaped valley A valley with a U-shaped cross-section. In their lower courses rivers may build up wide flood plains which can be like a very flattened U in cross-section. Better U-shaped valleys are those which have been cut by glaciers, where ice erosion has flattened the floor and steepened the sides of the pre-existing river valley; for example, in the Lake District or the Alps. Glacial U-shaped valleys are often quite straight as the old river valley spurs have been truncated by the glacier.

The floor will have been flattened, although in places it may have been over deepened and gouged out to form a hollow. The hollow will be elongated, along the valley, and if filled with water will form a ribbon lake. The post-glacial river flowing in a glaciated valley may be too small for the enlarged valley, and it will be called a misfit. Tributary streams may come rushing down the steep valley walls from hanging valleys high up on the hillsides.

V

valley An elongated depression which is lower than the surrounding countryside, and is usually the product of river erosion. It may be no more than a few metres deep, or as much as 5000 m, in the case of some Himalayan valleys. It may be just a few metres in width, or several kilometres; for example, the Thames, Mississippi, or Amazon. The cross-section of the valley may be a steep V, a more flattened V, or even a U, and in many cases it will be asymmetrical. The shape will depend on the rock type and the amount of river erosion. The valley will have been eroded by a river, and the erosion may continue for millions of years. Some valleys have lost their river because it has been captured, or has gone underground, and they are called dry valleys. There are many examples in chalk and limestone regions. In mountainous areas, valleys are young and tend to be narrow and steep, often with rocky beds. In lower areas the valleys widen out and become less steep. The meanders of the more mature rivers cause lateral erosion, which widens the valley. In lowland areas, where rivers and landscape are in their old age, the valley has a flat floor, is very wide, and has gentle sides.

vauclusian A spring which emerges from underground in a limestone area. It takes its name from Fontaine de Vaucluse, a large and powerful spring in southern France.

veering A meterological term denoting the clockwise change of direction of a wind, such as from southerly to westerly. See also **backing**.

vegetation The plant life of a particular area. It will depend on the rainfall and temperature, and may be affected by rock and soil type. Humans may remove the natural vegetation, and use the land for cultivated vegetation, such as grass or food crops.

vein A thin deposit of a mineral occupying a crack or crevice in the rock. After volcanic activity, heated liquids and gases spread out through cracks in many of the surrounding rocks. The fluids contain minerals, which solidify as they cool down. Some will be metallic ores and precious stones; the majority are likely to be less valuable minerals, such as quartz or felspar. Lead, silver, tin, zinc, gold, barytes, fluorspar, and many other minerals can be formed in this way, and they can be quarried or mined from veins, as in Devon, Cornwall and the Peak District.

veld A large expanse of grassland, especially in southern Africa. It is the result of the climatic conditions which are not wet enough to enable trees to grow. Much of the Transvaal in the Republic of South Africa is veld, although in places the grass has been ploughed up by farmers to grow wheat and maize.

vernal equinox see **equinox**

vertical corrasion Corrasion on a rocky riverbed which works downwards to produce potholes. Gradually the potholes become enlarged until they merge with their neighbours. In this way the bed of the stream will be lowered.

vertical exaggeration The exaggeration of the vertical scale in a cross-section in order to show undulations. Although the horizontal scale is generally the same as that of the map from which the section is being taken, the vertical scale is likely to be much bigger. For example, a horizontal scale of 2 cm per km square taken from a 1:50,000 map, will often be used with a vertical scale of 2 cm per 100 m, which will give a vertical exaggeration of 10 times.

vertical interval The height difference between neighbouring contours on ordnance survey maps. In the recent editions of 1:50,000 maps, the vertical interval is 10 m.

vertical link In manufacturing, one of a chain of sequential processes. In industrial organization it is best to have vertical links in the same factory. For example, in textile manufacturing the cloth is spun, then woven, and finally made into a finished item of clothing in the same factory. This is vertical organization. The old cotton industry of Lancashire used to have horizontal organization; the spinning took place in the south around Bolton, Oldham and Rochdale, the weaving was in the Blackburn and Burnley area, and the dyeing and finishing were carried out in Manchester.

viscous lava Acidic slow-flowing lava, which forms steep mountain sides. Basic lava is less viscous and can flow quite swiftly, forming gentle slopes.

viticulture The cultivation of grapevines. Grapes are mostly grown for making wine, but in places they are used for currants, raisins and sultanas; for example, in Greece, Turkey and Australia. As well as the traditional wine-producing countries of France, Italy and West Germany, many newer areas, such as California, Bulgaria and Australia are beginning to export their wines.

volcanic bomb see **pyroclastic rock**

volcanic plug A hard mass of solidified lava in the vent of a volcano. Over a period of millions of years, the volcano will be eroded, but the hard plug will survive as a small isolated hill in the crater. Also called **volcanic neck**. See also **puy**.

volcanic rock Any rock formed on the surface of the earth as a result of igneous activity. Because they form on the surface volcanic rocks cool quickly and consist of small crystals, invisible to the naked eye. They are mostly basic in chemical content and form basalt. If they are acidic they form rhyolite. If they cool very quickly, as for example in a submarine eruption, they form obsidian or volcanic glass.

volcano* An opening in the earth's crust, through which molten rock, ashes, steam, etc., are ejected. Some volcanoes erupt in an explosive way, throwing out rocks and ash; others are effusive, and lava flows out of the vent, and there are volcanoes which are both explosive and effusive. The vent or crater is usually at the top of the volcano, but occasionally volcanoes blow out of the side of the hill. Some volcanoes have built up large mountains as a result of many eruptions; for example Mt Etna, which has a circumference of over 120 km. Volcanoes may be extinct, which means they have not erupted for centuries; dormant, which means they have not erupted for several decades; or active, which means they have erupted in recent years. There are over 600 active volcanoes in the world. The largest number can be found around the Pacific Ocean in an area known as the fiery ring of the Pacific; that is, in Japan, Indonesia, Philippines, New Zealand, Chile and the western United States. The next most impor-

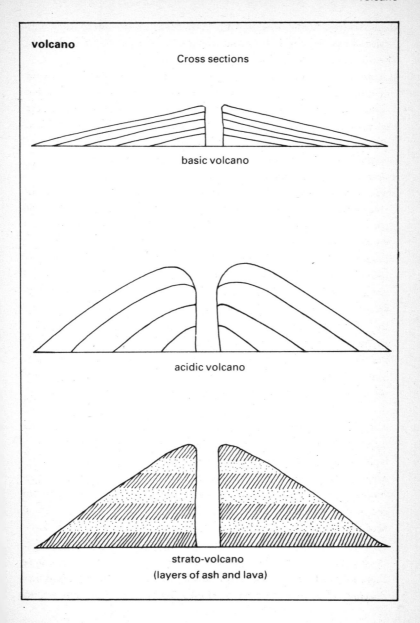

volcano

Cross sections

basic volcano

acidic volcano

strato-volcano
(layers of ash and lava)

tant area is the mid-world belt running from the West Indies and through the Mediterranean. In the centre of the Atlantic is the mid-Atlantic ridge, including Iceland, and there are a few minor areas; for example, in Antarctica and near the Great Rift Valley of East Africa.

There are different types of volcano according to the eruptions which caused them, and the shape of the mountain which has been formed.

1 Basic lava, **Hawaiian** type; for example, Mauna Loa. Free flowing runny lava has built up Mauna Loa into an enormous mountain. It is called a shield volcano because it is the shape of an upturned shield. Mauna Loa is over 9000 metres in height, and has grown up from the bed of the sea to form an island. It is a larger mountain than Mt Everest. Many basic eruptions come up through fissures, rather than through circular craters.

2 Acidic lava, **Peléan** type; for example, Mt Pelée in Martinique. The very viscous lava has formed a steep sided volcano. In the big eruption of 1902 a spine of lava was left protruding from the crater, although it was worn away by erosion during the next two years. A large nuée ardente rolled down the mountainside and killed all 28,000 people in the town of St Pierre at the foot of Mt Pelee.

3 **Strombolian** volcanoes (after Stromboli in the Lipari Islands, Sicily) are mildly explosive. Activity is quite mild but is frequent.

4 **Vulcanian** volcanoes (after Vulcano in the Lipari Islands) erupt slightly more explosively. Eruptions but less frequently than the Strombolian type. The lava is more viscous and there are dark clouds and a great deal of steam.

5 **Vesuvian** volcanoes are even more explosive. They expel cinders and ash, as well as pouring out lava. Vesuvian volcanoes may be dormant for long periods, and then have a massive eruption. The lava is generally acidic.

Strombolian, Vulcanian and Vesuvian eruptions all gave rise to composite volcanoes, as the cone-shaped mountains consist of layers of ash, cinders and lava.

6 The **Krakatoan** type is very violent, but may not give off any lava.

Most volcanoes are situated on or near the margins of the plates which make up the earth's crust. Where the plates are moving apart for example, below Iceland, the volcanoes erupt basic lavas. Where the plates are moving together, the volcanoes tend to be more explosive, and they erupt acidic lava, together with cinders and ash; for example, Japan, Philippines, Indonesia and Mt St Helens in the western United States.

Von Thunen, J.H. A 19th century German estate-owner, who lived near Rostock in eastern Germany. In 1826 he wrote a book called *Isolated State*, in which he outlined his ideas about the economics of land use. Using a farm or a village as a central place, he believed that the most intensive farming, dairying, and market gardening, would take place in a concentric zone near to the central place. Further away from the central place, there would be another concentric ring of less intensive types of farming, because it would take the farmer longer to walk out to the fields. In this area he would grow crops which required little attention. Beyond the second ring, there would be a third concentric ring of very extensive farming, where the land, crops and animals needed very little time spending on them. Von Thunen's ideas of location theory in agricultural land use were based on the assumption that the land was flat and featureless, and that transport facilities would be equal in all directions. This is called an isotropic surface and does not really exist, though flat reclaimed land, as in the polders or the Fens, is nearly isotropic. He also assumed that there would be only one form of transport and one market. However his general ideas still have some relevance, on various scales and in various locations, all over the world. Around farms and villages in many Third World countries and in southern Italy, the distinctive zones of decreasing intensity can be seen clearly. On a larger scale, the same kind of zoning can be seen in European farming, as the most intensive areas are in the centre, near London, Paris, Brussels, and less intensive areas can be found further away from the major towns.

V-shaped A valley with a V-shaped cross-section formed by river erosion. Both vertical and lateral corrasion take place.

vulcanicity The process connected with the formation of and the movement below and through the earth's crust of magma and other associated materials.

W

wadi A narrow steep-sided valley found in deserts and semi-arid regions. Wadis are dry valleys for much of the year. Rainfall is not frequent in desert areas, but such rainfall as there is falls in heavy showers. As much of the land is bare with no vegetation, surface run-off can be rapid. The water will carry sand and small rock fragments, and so a great deal of erosion takes place. Vertical corrasion can form narrow steep-sided valleys, which can grow to several hundred metres in depth. In humid lands there is more frequent rainfall and rainwash to wear away the valley sides making them more V-shaped in cross-section. Some wadis extend for long distances, and it is possible for a heavy storm in one location to cause flooding in a wadi which can rush many kilometres, possibly into areas where there has been no rainfall. Such rapid and short-lived floods have been known to catch travellers unawares. Wadis are common in Egypt, Libya, and many other desert areas. Also called **arroyo**.

warm front A leading edge of warm air. The warm air will be rising gradually over the colder air which has gone ahead. As the air rises it will cool and form cloud, probably cumulus or cumulonimbus. Eventually it may give rise to some rain, which will be frontal rainfall. A depression often contains a warm front, and it is the first front to arrive. It will be followed a few hours later by a cold front. In the Northern Hemisphere the warm front is generally located to the south of the centre of the depression. See also **cold front, occluded front**.

warm sector A region of mild or warm air situated between the warm and cold fronts in a depression. It is generally cloudy with stratus, and may give some light rain or drizzle, but is often dry. It lasts for only a few hours before the cold front arrives to give heavier rainfall and lower temperatures. In most depressions the cold front gradually catches up and merges with the warm front to form an occluded front; the warm sector will then have disappeared.

water cycle   see **hydrological cycle**

waterfall   A steep or vertical descent of the water of a stream or river. Where there is a change of rock type on the bed of a stream, there are different rates of erosion. This creates an irregularity on the long profile of the stream and forms rapids or a waterfall. The softer rock is eroded more quickly causing an abrupt change in the gradient. If there is a stratum of hard rock which is vertical or nearly so, only a small waterfall will be created. If the harder rock is horizontal, a much larger fall can form. The softer rocks around may be worn away and may possibly undercut the hard rock. If this happens, there will be a collapse, though the waterfall will not disappear. It will just move upstream, and gradually a gorge will form downstream from the waterfall. Examples include the Niagara Falls and High Force on the River Tees.

water gap   A gap in a ridge cut by a stream or river. There are several examples on the chalk hills of southern England, such as the gaps cut by the Wey and the Mole through the North Downs. A famous gap in the United States is the Delaware Water Gap, where the River Delaware has cut through the sandstone of the Kittatinny Mountains.

water power   The power produced by the movement of a body of water. For example, a stream can turn a water wheel, which can drive a machine. Water power was used in the early cotton and woollen industries of Lancashire, Yorkshire and New England. It has been used for centuries in mills which grind wheat into flour. In many parts of the world, water from rivers is used to drive turbines and generate electricity in hydro-electric power stations. There are several in North Wales and the highlands of Scotland, and hydro-electric power is the most important type of electricity in some countries; for example Norway, Sweden and Switzerland. There are large schemes in countries all over the world; for example, Iguassu in Brazil and near Bratsk in the Soviet Union.

watershed   The high ground which forms the boundary or dividing line between two river basins. On one side of the boundary line the water will drain in one direction; on the other side of the boundary line it will drain in the opposite direction. The watershed is normally a ridge or a piece of ground that is higher than the surrounding areas. It is often a very irregular line. Also called **water divide, water parting**.

waterspout A funnel-shaped mass of water, which is similar in formation to a tornado. It occurs over water when a heated patch of air rises, and whirls around in a circular or corkscrew fashion. Waterspouts are mostly seen in tropical areas, especially in late summer, but there are occasionally waterspouts off the coast of England. When waterspouts go onto land they become tornadoes, and when tornadoes go over lakes or the sea, they become waterspouts. Waterspouts are generally less than 30 m in height, and they last for half an hour or so. Often a few occur on the same day and they may be accompanied by strong gusts of wind. Large boats can sail through them undamaged, but small sailing vessels would be knocked over.

water-table The level below which the rock is saturated. After rain has fallen, the water gradually trickles downwards through the pores in the rocks. The level at which all the pores are full is the water-table. The height of the water-table gradually changes, moving up or down depending on the recent rainfall. After heavy rain the water-table will rise, and in spells of wet weather it could reach the surface, causing puddles or lakes to form. On average the water-table lies a few metres below the surface, although, in the Fens and polders it can never be more than a few centimetres below the surface, and in upland limestone areas it can be as much as 100 m down. The line of a water-table generally follows the shape of the land surface, although it is normally neither as steep nor as high. Water located below the water-table is called ground water.

wave-cut platform An outcrop of level or gently sloping rocks on the shore, which have been scraped and smoothed by abrasion. The sea carries sand and pebbles, which scrape away at any exposed rock surfaces. A wave-cut platform is often created at the foot of cliffs and may gradually increase in size as the cliffs retreat inland because of erosion. Also called **abrasion platform**.

wave refraction The bending of a wave as it approaches the shore, due to a variation in the depth of the water. If the waves are coming in diagonally to the coastline, the part of the wave in shallower water will slow down, and the crest line of the wave will begin to curve. Waves also bend near headlands because of the same effect caused by shallowing water.

weather The meteorological conditions which are being experienced over a short time period. When the weather is recorded over a few days and then averaged out, it can be called climate. The weather is the result of the state of the atmosphere, and the pressure, wind, temperature, humidity and rainfall caused by the atmospheric conditions. In Britain there are frequent and rapid changes of weather. Near the equator the daily weather is much the same, day after day, and so the weather is almost the same as the climate. In deserts and near the poles, the daily weather may be similar for a few days or even weeks.

weathering The decay and disintegration of rocks caused by the effects of weather and atmosphere. Weathering is one of the major processes involved in denudation and causs the breakdown of rocks in situ. No movement or transport of the weathered material is involved; otherwise, the process would be called erosion. Weathering may be mechanical, chemical, or biological. **Biological weathering** can be the result of roots opening up cracks in the rocks, and worms, rabbits, bacteria, etc. can also contribute to rock disintegration. **Chemical weathering** may take several forms. Rainwater contains carbon dioxide from the atmosphere and can cause solution of salts in rocks, or carbonation in chalk and limestone. Oxidation is the result of rainwater and oxygen from the air leading to disintegration of ferrous minerals. Hydration is another type of chemical weathering in which a mineral combines with water. **Mechanical weathering** (also called **physical weathering**) is mainly the result of heating and cooling. The heat of the sun causes rocks to expand and when they cool at night they contract. The expansion and contraction of minerals causes rocks to crack. If the temperatures fall below freezing point, the process is more effective. Any water in the rocks will freeze and expand by 9% of its volume. Alternate freezing and thawing will lead to the disintegration of any rock. See also **exfoliation**.

Weber, Alfred Weber described a standard theory of industrial location. He believed that the optimum location for an industry would be where transport costs were lowest. Collecting raw materials and distributing the finished product would be the vital consideration. In the case of the steel industry, the best location would be somewhere between the sources of iron, coal and limestone. The availability of a labour supply and the location of markets would also be important. The ideas in Weber's locational model are still economically sound, though conditions have changed appreciably since he wrote in 1919. In the case of the steel industry, most new works are so large that no one place can provide all the raw materials. Because of this, coastal locations are generally used so that large quantities of materials can be transported by sea, the cheapest form of bulk transport. In many other industries transport costs may be quite low, especially if few materials are required and the finished product is quite small.

wedge An area of high pressure extending from an anticyclone, similar to but narrower than a ridge of high pressure.

westerlies The major planetary winds in temperate latitudes. There are two main types of large scale planetary winds, which blow out from the tropical high pressure regions; they are the westerlies and the tradewinds. The westerlies blow towards the poles from the horse latitudes, but because of the rotation of the earth, become south-westerlies in the Northern Hemisphere; for example, near Britain, and north-westerlies in the Southern Hemisphere, for example, near Tasmania or southern Chile. While the trades are quite persistent, the westerlies are very variable. Most of the weather in temperate latitudes comes from the west, but it includes depressions and anticyclones, in which the winds will be blowing in different directions. The effects of the continents and oceans interferes with the flow of westerly weather in the Northern Hemisphere, but there is less interference in the Southern Hemisphere, which is mainly ocean. In the Northern Hemisphere, westerly weather affects Britain, Iceland and Norway for much of the year, and in winter, when the world's winds all move southwards, westerlies can affect the Mediterranean lands, too, bringing winter rainfall.

wet bulb thermometer see **hygrometer**

wet day A day on which measurable rain has been recorded.

wet site The site of a settlement where water is available in an otherwise dry area, such as a well in a desert, or a spring in a limestone or chalk area.

wet spell (According to the Meteorological Office) a period of at least 15 consecutive days on which 1 mm of precipitation has been recorded.

wheat A cereal grass of the genus *Triticum*, which is widely cultivated in temperate areas. It grows best in areas with more than 120 frost-free days, where temperature are 17° to 20°C in the last two months, with a dry spell for harvesting. Total rainfall requirements are 400 to 700 mm depending on the season of fall and the amount of evaporation. Major growing areas include the prairies and the steppes. Wheat is also grown in East Anglia, where there are large fields and highly mechanized farms. It is the major cereal in temperate latitudes.

whirlpool A circular movement of water in a river or the sea, caused by the meeting of two currents or by the shape of the rocks on the bed of the river or sea.

whirlwind A whirling column of air caused by a small rising current of hot air, which may pick up dust and cause a dust storm.

white-collar worker A non-manual worker who works in a clean environment, such as in an office. It is a term also used to describe professional people. See also **blue-collar worker**.

white ice Ice on or near the surface which still contains some air. At greater depths all the air is removed by compression, and the ice will look blue. As the ice changes from white to blue, there is melting followed by freezing, which is called regelation.

wind The movement of air from one place to another. Air moves from high pressure to low pressure in an attempt to balance out the pressure differences. As the wind travels it is diverted to its right in the Northern Hemisphere or to its left in the Southern Hemisphere. If the isobars are close together, the pressure gradient will be steep, and the wind will be strong. More widely spaced isobars mean there is a low pressure gradient, and the wind will be gentle. Winds are named by the direction they are blowing from, so a south-westerly wind actually blows towards the north-east. Winds are recorded by anemometers and can be measured in knots, miles per hour, or metres per second. The Beaufort Scale is a traditional method of measuring wind speed.

windbreak A line of trees planted to provide shelter from strong or persistent winds. Many houses in the Rhône valley are on the south side of clumps of trees, which protect them from the mistral. Farms and other isolated buildings in the Pennines often have a few sheltering trees to break the full force of the wind. On the prairies and the pampas rows of trees are frequently planted round the farmhouses. Crops in the Rhône valley are sheltered by windbreaks of trees, but in the Scilly Isles windbreaks may be hedges. Walls are also important for shelter on many Pennine farms, and also in the Hebrides. Some windbreaks are planted to protect soil and prevent erosion.

wind chill The perceived cooling effect of wind in low temperature conditions. The air temperature feels much colder if there is a wind blowing. The physiological temperature is the way in which the human body feels the temperature of the air, and this is certainly strongly influenced by wind speed and direction. For example, in Britain an easterly wind in January will feel far worse than a south-westerly wind of the same speed. Temperatures below zero feel very cold if the wind speed is 10 km per hr or more, and if the wind reaches 40 km per hr temperatures of plus 5°C will also feel very cold.

wind gap A gap, col, or small notch in a hilltop, through which the wind can whistle. The gap would have been carved by water erosion, possibly from a glacial melt-water stream. A wind gap may be a dry valley in chalk or oolitic limestone landscapes, or may be the result of river capture; it will not have been formed by the wind.

wind rose A diagram used for showing wind direction at a weather station. The eight principal points of the compass are shown, and the length of line on the rose indicates the number of days the wind has blown from a particular direction. The number of calm days is generally written in the centre of the rose.

windward A term used to denote the side of a hill or building which faces into the wind. The opposite is leeward.

winterbourne A stream which flows in winter but dries up in summer when the water-table is lower. Such streams are found mainly in chalk areas and the name is used only in southern England.

winter solstice December 22nd. See also **solstice**.

wold A chalk hill in Lincolnshire or Yorkshire.

wrench fault see **tear fault**

X

xerophyte Any plant which is adapted to survive in arid or semi-arid conditions. Some xerophytes have long roots to reach underground supplies of water; others have small or thick leaves, or no leaves at all. Some have fleshy stems in which to store water, and many have a thick bark or a waxy layer to reduce transpiration. Xerophytic plants such as cacti are found in deserts, and thorny bushes such as sage brush are found in arid and semi-arid regions. Palm trees survive in many places, and tough tussock grasses occur widely. Many flowers grow in deserts after a rainstorm, and the plants go through a rapid life cycle. When the plants die, seeds are left to lie dormant until the next rain falls.

Y

yardang A narrow steep-sided ridge found in various arid areas; for example, Turkestan in the southern Soviet Union. Yardangs often occur in groups with several ridges running parallel to each other and oriented in the direction of the prevailing wind. They are the result of sand corrasion. Wind blowing between the yardangs removes any depostional material which is accumulating.

young mountain A recently formed mountain. The young mountains include the highest mountains in the world. Fold mountains formed by alpine earth movements have undergone only 40 or 50 million years' erosion and, on a geological time scale, this makes them a young landscape feature. Although the alpine folding created several large mountain ranges, there were a few smaller hills formed around the edges of the main ranges, examples of which are to be found in Britain and include the Weald, the Isle of Wight hills and the small hill on which Windsor Castle stands.

youthful 1 (Of a river valley) in its upper or mountain course. The valley is narrow and V–shaped in cross-section and active vertical erosion will be taking place. 2 (Of a landscape) still covered with mountains, for example, the Himalayas or Alps. As the landscape is eroded, the mountains become lower and the valleys wider, and the landscape gradually becomes mature or middle-aged.

Z

zero poplulation growth The point at which there is no increase in population. Zero population growth occurs when the birth rate and death rate are approximately the same. It has been reached by Sweden, France and Britain. When a country has an optimum population, it will hope to achieve zero population growth. Many Third World countries have a population increase of 3% per annum, and the world average is about 2%. Accordingly, there is much to be done in order to achieve zero population growth throughout the world. Some countries, such as Australia or New Zealand, still receive immigrants, and this will maintain an increase in population even when the birth rate and death rate are roughly similar.

zeugen A pedestal rock with a tabular-shaped profile caused by wind erosion. Variations in resistance to erosion are responsible for the irregular profile, but there is normally greatest erosion just above ground level, where wind and sand activity are at their maximum. Examples can be seen in most deserts, but especially in the southern Soviet Union. There are similar rocks, though eroded to a lesser extent, on the millstone grit outcrops of the Pennines. See also **pedestal rock**.

zonal soil A soil found in large areas having a similar climate and vegetation. For example, climatic conditions which are ideal for grass in the prairies and steppes have contributed to the great accumulation of humus which created the chernozem soils. The taiga of Norway, Sweden, Finland, the Soviet Union and Canada has created an acidic soil called podsol. Zonal soils fall into one of two main groups: pedalfers and pedocals.